abc's

of

Citizens Band Radio

by

Len Buckwalter

HOWARD W. SAMS & CO., INC.
THE BOBBS-MERRILL CO., INC.
INDIANAPOLIS • KANSAS CITY • NEW YORK

THIRD EDITION

SEVENTH PRINTING—1976

Copyright © 1962, 1964, 1966, 1973, and 1975 by Howard W. Sams & Co., Inc., Indianapolis, Indiana 46268. Printed in the United States of America.

All rights reserved. Reproduction or use, without express permission, of editorial or pictorial content, in any manner, is prohibited. No patent liability is assumed with respect to the use of the information contained herein. While every precaution has been taken in the preparation of this book, the publisher assumes no responsibility for errors or omissions. Neither is any liability assumed for damages resulting from the use of the information contained herein.

International Standard Book Number: 0-672-21021-5
Library of Congress Catalog Card Number: 73-85139

Preface

Following approximately 20 years of steady development in the field of radiocommunications, the Federal Communications Commission, in 1958, allocated a communications band for general use by citizens of the United States. Called the Citizens Radio Service, this band of channels is provided for applications which do not fit in the established categories of municipal, emergency, and transportation services. The Citizens band is especially for "business and personal" communications.

To derive maximum benefit from the CB service, a user must have an understanding of the rules, regulations, procedures, and equipment operation. This book is written to supply such information. The applications of CB radio are discussed, and the few restrictions imposed by the FCC are explained. This is a guide to the establishment of a CB system—from the initial "spark" of an idea to the completed and functioning installation.

In the Preface to the First Edition of this book it was stated that "by following the directions in this book and referring to the discussions, anyone should be able to set up an efficient CB communications system." The letters and comments I have received indicate many readers of the first two editions have done just that. As a result this book has been updated to include changes in rules and regulations and to reflect the improvements that have been made in CB equipment.

LEN BUCKWALTER

Contents

CHAPTER 7

CHAPTER 8

CHAPTER 9

CHAPTER 10

CHAPTER 11

CHAPTER 12

APPENDIX

1

Introduction

Following 20 years of phenomenal growth in the field of radio communications, the Federal Communications Commission (FCC) in September 1958 allocated a two-way radio service for use by the general public. The miracle of instant wireless communications, previously reserved for a select few, became available to virtually anyone who wanted to set up his own radiotelephone system. Compared with other radio services, licensing for the Citizens band is extremely simple—just fill out a form and mail it to the FCC secretary. There is no code test, and no technical skill or specialized knowledge of radio required.

APPLICATIONS OF CB RADIO

Less than 10 years after Citizens band radio (CB) came into existence, it became the largest, single, two-way radio service the world had ever known. Nearly a million American citizens obtained licenses and reaped the benefits of mobile communications (Fig. 1-1) in their daily lives. A follow-up FCC study revealed that about 80% of all CB license holders installed sets in both car and fixed station (home or office). While 13% of CB sets went into boats, 7% were installed in farm vehicles. Pilots (Fig.

1-2) joined the rush to CB by installing sets aboard some 8000 light aircraft. About 29% of these license holders also operate "handie-talkies." Fig. 1-3 shows a system of one fixed (base) station and four mobile units.

The history of mobile radio is one of strict electronic standards and professional consultations. Installation and servicing must be done under the supervision of an FCC-licensed technician. Annual checks of on-frequency operation are another official requirement that boosts the cost of owning and maintaining two-way radio. CB has few such limits, which accounts for part of its popularity. If a person is mechanically inclined and has a few simple tools, he can install a CB set in his car, home, or boat, erect an antenna, and start operations. Only when trouble develops within the transmitter circuits does a CBer have to seek the services of a qualified repairman.

Personal Uses

Personal communication, estimated at 70% of all activity, is the most popular application of CB. The classic example is the "late-for-dinner" call. If you are stuck in

Fig. 1-1. A typical mobile CB set installed in automobile.

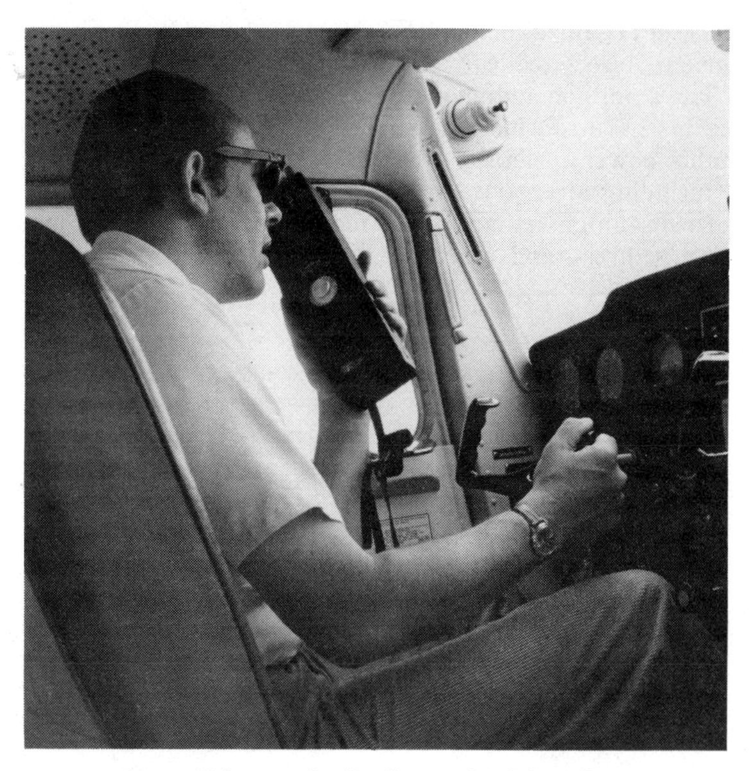

Fig. 1-2. High-power "handie-talkie" used in light airplane.

traffic and the roast will burn before you can get home, pick up the microphone and tell someone at home about the delay. About 1 in every 39 cars on the road was equipped with CB by the early 1970s. The medium is so widely used on camper vehicles that manufacturers produce special antennas for this application alone. Range over water is good enough (up to 30 miles) for small-boat owners to choose CB over costly vhf-fm marine radio, even though CB is not officially monitored by the Coast Guard. Hunting (Fig. 1-4) and other outdoor activities are made safer by tiny, but powerful, portables.

REACT

It is estimated that 1 out of every 10 CBers is somehow connected with a volunteer or public service. One of the largest organizations is REACT (Fig. 1-5) (*R*adio *E*mergency *A*ssociated *C*itizens *T*eams) which reports more than a million automobile emergencies per year. Many teams have considerable emergency equipment, like portable power generators, inhalators, and special vehicles (including aircraft). REACT teams respond to a variety of calls—highway accidents, mechanical breakdown, out-of-gas, fires—and relay the messages to the proper au-

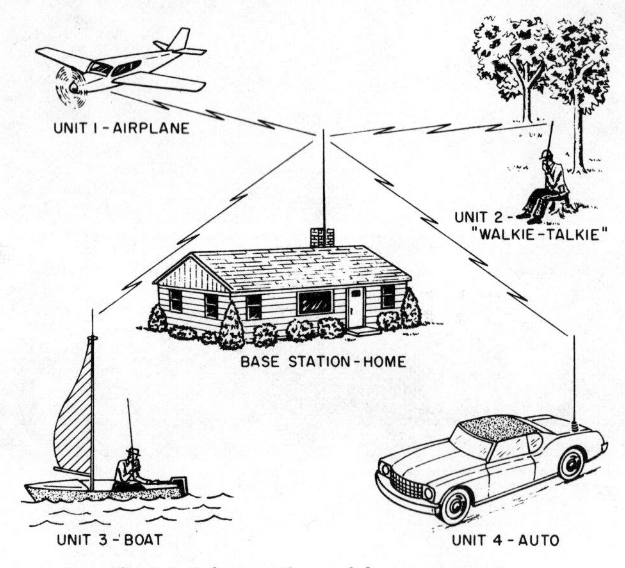

Fig. 1-3. A base station and four remote units.

thority. About 2000 REACT teams are engaged in such activity and 35% of them are officially affiliated with civil defense, police, or fire authorities in their communities. The degree of CB assistance actually runs much higher. The FCC has estimated that 5 million emergencies (Fig. 1-6) of all types have been reported within one year via CB radio.

Business Uses

While only 30% of the CB activity is devoted to business, it is easy to see that business uses of CB can result in a more profitable operation. The efficient dispatching

Fig. 1-4. Hunter using CB in the field.

Fig. 1-5. REACT is one of the largest organizations of CB volunteers.

of cars and trucks en route can more than make up the cost of the equipment and its servicing within a short time by reducing the amount of "deadheading." Businessmen, professional people, and salesmen can be in constant contact with their offices—quickly informed of important messages. CB proves indispensable to the large-scale operations often encountered in service stations (Fig. 1-7), farming, construction projects, plant security (Fig. 1-8), and long-haul trucking.

Fig. 1-6. A call for help on CB.

Restrictions

Although the restrictions on the use of the band are reasonable, it should be remembered that it is set up on a "party-line" basis. You will, on occasion, have to wait until another station has completed its transmission and the channel is clear. Keeping transmission time down to the shortest possible length (as indicated by FCC law) is essential to keeping the Citizens band an efficient medium.

This band is certainly no place for the type of activity reserved for amateur radio where lengthy discussions of equipment, antennas, etc., are commonplace and legal. Contacting another station merely to see how far your signal "gets out" is a practice which prompts increased government regulation. Soon after the initial rulings on CB were published, the FCC followed up with several amendments to curb some of the abuses of the band. Many of them apparently were due to misunderstanding of the role of CB. Another limit to CB radio is the limited range. The band, by definition, is a short-range service. A power maximum of five watts and regulated antenna height tend

Fig. 1-7. CB being used in service station.

to keep the range to 10 to 15 miles. The service is not an end in itself (i.e., a hobby or form of recreation), but rather a medium for the brief exchange of definite messages.

CLASS-D BAND

This book is devoted solely to the class-D band, where virtually all CB activity occurs. It was created from a former 11-meter amateur band (on 27 MHz). There are other Citizens bands (Table 1-1), but they have only limited or specialized value. One is in the ultrahigh frequency (uhf) region of approximately 460 to 470 MHz, where equipment is costly and signals unreliable unless handled through elaborate antenna systems mounted in high locations. Another Citizens band, class C, is for radio control of remote objects, an entirely different subject that will not be covered in this book.

Courtesy Raytheon Co.

Fig. 1-8. "Handie-talkie" provides three-mile range for security guard.

Besides proving immensely popular, the class-D band has enjoyed remarkable technical advancement. In 1958 a typical CB set had vacuum tubes, a weight of about 10 pounds, and one channel for communication. In less than 10 years, weight and size were slashed by the use of semiconductors, and the number of channels in many sets

Table 1-1. The Three Bands for CB

Class	Frequency Range (MHz)	Power (Maximum Watts Input)	Authorized Emissions
A	462.55 - 462.725	60	8A2, 8A3, 20F2, 20F3 (primarily voice)
	467.55 - 467.725 (8 frequency pairs)		
C*	26.995 - 27.555 (6 channels)	5 (30 watts on 27.255 MHz only)	0.1A1, 8A2 (controls signals only)
	72 - 76 (5 channels)	1	0.1A1, 8A2 for the radio control of model aircraft only
D	26.965 - 27.225 & 27.255 (23 channels)	5	8A3 (voice only)

*Class C is for the radio control of remote objects or devices such as model airplanes and garage door openers. Voice communications prohibited.

went to 23 (Fig. 1-9). Numerous sets today use noise- and interference-rejecting circuits that were once found only in commercial equipment costing 10 times as much. Single sideband, a highly advanced transmission method (described in Chapter 5), is available from several different manufacturers (Fig. 1-10). Elaborate base sta-

Courtesy Browning Laboratories, Inc.

Fig. 1-9. A small 23-channel solid-state transceiver.

Pearce-Simpson, Inc.

Fig. 1-10. A CB single-sideband transceiver.

tions have built-in test equipment and accessories that instantly indicate trouble and improve operating convenience. The variety and size of equipment is adaptable to almost any purpose—home, office, automobile, marine, portable, or pocket. Citizens band has, in fact, emerged as the "two-way radio for everyone."

2

Licensing and Eligibility

Almost any United States resident over the age of 18* may file for a CB radio license. Actually, the minimum age limit is not quite that restrictive. The license is for the station equipment only; there is no need for an operator's ticket. Thus, a person under 18 may use a CB system, if the licensee assumes responsibility for proper operation. The responsibility applies to anyone you authorize to participate in your private communications setup. A single station license covers all units under your control.

Provisions for licensing partnerships, corporations, and state and local governments also appear in the regulations. The notable exception in the case of group licensing occurs where any corporation is controlled by a foreign government.

The eligibility requirements for CB are intended to give the largest possible group access to the benefits of two-way radio. But, this does not imply that anyone may indiscriminately use the band to expedite an illegal business. A CB station may not be operated for any purpose contrary to federal, state, or local law.

*At the time of this printing, the FCC has proposed a 16-year-old minimum age.

Since there are no examinations required for the license, all paper work may be completed at home and the filing done through the mail. The license fee is $4 and the license is good for 5 years. The only other cost involved is for a copy or copies of the regulations.

FCC RULES AND REGULATIONS

The first step toward obtaining a license is to secure a current copy of Part 95 of the FCC Rules and Regulations. According to FCC ruling this must be in your possession.

The laws of telecommunications are grouped into 10 volumes which cover all the services under FCC jurisdiction. Part 95, dealing with CB radio, appears in Volume VI which also includes parts on Amateur and Disaster Communications and costs $5.35. This one-time charge entitles the buyer to receive all future amendments to Volume VI for an indefinite period. On receipt of an amendment, the obsolete page is removed and the new one is inserted (Fig. 2-1); thus, continuity is not disturbed.

Volume VI is available by writing to:

Superintendent of Documents
U.S. Government Printing Office
Washington, DC 20402

Enclose a check or money order, and allow about two weeks for delivery.

(Part 95 covers not only the legal aspect of CB, but constitutes a sort of "handbook" of operating procedures, frequency allocations, and a wealth of other essential information.)

HOW TO FILL OUT AND FILE APPLICATIONS

The application form for the CB license is secured from your local FCC field office. (Select the nearest one from the list given in the appendix.) Request by telephone or letter FCC Form No. 505.

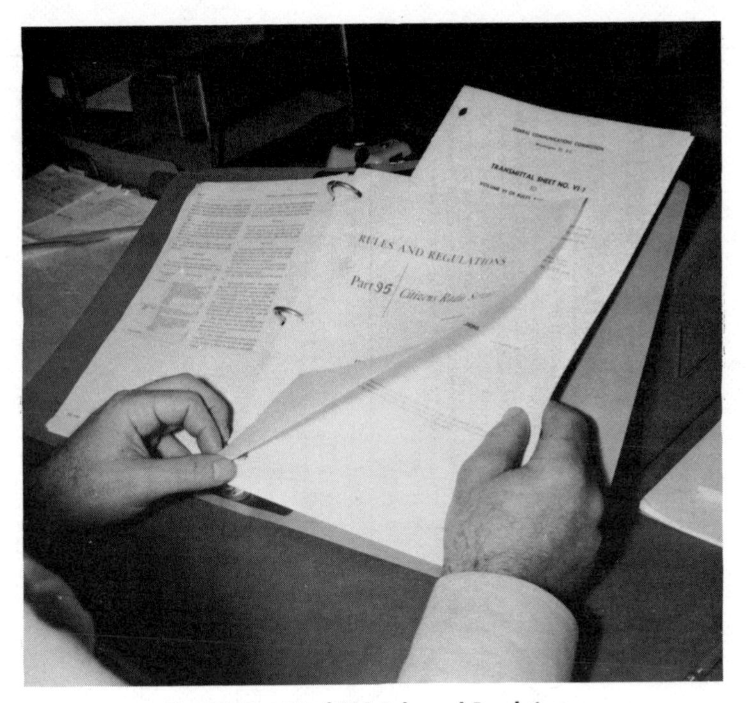

Fig. 2-1. Part 95 of FCC Rules and Regulations.

The CB license application Form 505 is shown in Fig. 2-2. Although the application was once a complicated technical and legal document, it is now extremely simple and self-explanatory. Enter each letter of your name and other required information in the boxes provided in Items 1 through 4. Your date of birth, required in Item 2, is used to determine if you fulfill the age requirement. If your mailing address is different from the location of the equipment, fill in Items 5A, B, and C. The FCC must know transmitter locations for possible measurement or inspection.

Questions 6 through 9 determine whether you're an individual, a business, or some other entity. They also ask whether you want a class-C or -D license. In nearly all cases, class D is chosen because it's the popular voice band used by most CBers. The class-C band is for remote-control signalling devices. When answering Item 14, you'll

United States of America
Federal Communications Commission

Form Approved
GAO No. B-180227(R01 02)

FCC FORM 505

August 1975

APPLICATION FOR CLASS C OR D STATION LICENSE IN THE CITIZENS RADIO SERVICE

INSTRUCTIONS

A. Print clearly in capital letters or use a typewriter. Put one letter or number per box. Skip a box where a space would normally appear.

B. **Enclose appropriate fee with application. Make check or money order** payable to Federal Communications Commission. DO NOT SEND CASH. No fee is required of governmental entities. For additional fee details see FCC Form 76-K, or Subpart G of Part 1 of the FCC Rules and Regulations, or you may call any FCC Field Office.

C. **Mail application to Federal Communications Commission, P.O. Box 1010, Gettysburg, Pa. 17325**

NOTICE TO INDIVIDUALS REQUIRED BY PRIVACY ACT OF 1974

Sections 301, 303 and 308 of the Communications Act of 1934 and any amendments thereto (licensing powers) authorize the FCC to request the information on this application. The purpose of the information is to determine your eligibility for a license. The information will be used by FCC staff to evaluate the application, to determine station location, to provide information for enforcement and rulemaking proceedings and to maintain a current inventory of licensees. No license can be granted unless all information requested is provided.

1. Complete **ONLY** if license is for an Individual or Individual Doing Business AS

FIRST NAME INIT LAST NAME

2. DATE OF BIRTH

MONTH DAY YEAR

3. Complete **ONLY** if license is for a business, an organization, or Individual Doing Business AS

NAME OF BUSINESS OR ORGANIZATION

4. Mailing Address

4A. NUMBER AND STREET

4B. CITY

4C. STATE

4D. ZIP CODE

NOTE:
Do not operate until you have your own license. Use of any call sign not your own is prohibited

Fig. 2-2. A CB license

5. If you gave a P.O. Box No., RFD No., or General Delivery in Item 4A, you must also answer items 5A, 5B, and 5C.

5A. *NUMBER AND STREET WHERE YOU OR YOUR PRINCIPLE STATION CAN BE FOUND*
(If your location can not be described by number and street, give other description, such as, on RT. 2, 3 mi., north of York.)

5B. CITY

5C. STATE

6. *Type of Applicant* **(Check Only One Box)**

- ☐ *Individual*
- ☐ *Association*
- ☐ *Corporation*
- ☐ *Business Partnership*
- ☐ *Governmental Entity*
- ☐ *Sole Proprietor or Individual/Doing Business As*
- ☐ *Other (Specify)* __________

7. *This application is for*

- ☐ *New License*
- ☐ *Renewal*
- ☐ *Increase in Number of Transmitters*

IMPORTANT
Give Official FCC Call Sign

8. *This application is for* **(Check Only One Box)**

- ☐ *Class C Station License (NON-VOICE—REMOTE CONTROL OF MODELS)*
- ☐ *Class D Station License (VOICE)*

9. *Indicate number of transmitters applicant will operate during the five year license period* **(Check Only One Box)**

- ☐ *1 to 5*
- ☐ *6 to 15*
- ☐ *16 or more (Specify No.* ____ *and attach statement justifying need.)*

10. ***Certification*** I certify that:

- The applicant is not a foreign government or a representative thereof.
- The applicant has or has ordered a current copy of Part 95 of the Commission's rules governing the Citizens Radio Service. See reverse side for ordering information.
- The applicant will operate his transmitter in full compliance with the applicable law and current rules of the FCC and that his station will not be used for any purpose contrary to Federal, State, or local law or with greater power than authorized.
- The applicant waives any claim against the regulatory power of the United States relative to the use of a particular frequency or the use of the medium of transmission of radio waves because of any such previous use, whether licensed or unlicensed.

WILLFUL FALSE STATEMENTS MADE ON THIS FORM OR ATTACHMENTS ARE PUNISHABLE BY FINE AND IMPRISONMENT. U.S. CODE, TITLE 18, SECTION 1001.

11. __________

Signature of: Individual applicant, partner, or authorized person on behalf of a governmental entity, or an officer or a corporation or association

12. Date __________

application form.

state how many transmitters you plan to operate. If your answer is 16 or more, you'll have to explain the reason for the need (running a fleet of trucks, for example).

The last section of the form is a certification which you must sign. You'll agree that you'll operate under FCC regulations and not violate Federal, State or local law. You will say that you're not a foreign government or a representative of one. This does not mean, however, that you must be a U.S. citizen. The FCC will grant a license if you are a temporary or permanent resident of the U.S., so long as you have a U.S. mailing address. You must, however be acting as a private person, and not as a representative of a foreign government.

You don't have to notarize the application (an obsolete requirement) but you can be held accountable for false statements by reason of your signature.

After the application is signed, mail it with a check or money order for $4 to:

Federal Communications Commission
P.O. Box 1010
Gettysburg, Pennsylvania 17325

The processing, assignment of call letters, and delivery of your license will take from three to eight weeks. *Do not* attempt to operate your equipment until the actual license is received—such operation is strictly illegal. Do not attempt to speed up your license by writing to the FCC and inquiring about your application. This will usually slow it up instead because your application will have to be removed from the regular flow channel for checking. Thousands of applications are received each month; each is processed as soon as modern data processing equipment can handle it.

3

Choosing Equipment

The prospective buyer of CB equipment is confronted with an array of models produced by dozens of manufacturers. The products of individual companies are varied to suit the needs of diverse applications, but regardless of shape or style, each unit sold for class-D CB radio has to conform to FCC specifications. This is ensured by a rule that states the manufacturer must certify to the purchaser that the unit "type accepted" is in accordance with these regulations. Watch for this in advertising literature and instruction manuals.

LEGAL REQUIREMENTS

The buyer should be wary of equipment offered on the surplus market and labeled for CB use. True, the frequency coverage often falls within the 27-MHz region, but it takes a skilled technician to perform the necessary modifications. This expense often offsets the initial low cost. Another limitation on this type gear is the commercial radiotelephone license needed to perform certain changes and adjustments.

The safest course for the person with a limited electronic background is to select equipment from one of the reputable companies engaged in the manufacture of CB equipment.

Foremost among the FCC specifications covering class-D sets are crystal control, frequency tolerance, and transmitter power. The licensee is not *required* to understand these terms, but a basic understanding will promote intelligent selection and operation of a station.

Citizens Radio equipment offered in kit form must conform to the same regulations as the factory-assembled units. To ensure against off-frequency operation, the kit manufacturer must include in his kit a prewired and pretuned crystal oscillator. This "packaged" circuit contains the critical frequency-determining section of the transmitter. Fortunately, this represents about a fifth of the components in the kit and leaves most of the construction in the hands of the builder. This arrangement permits the benefits of the do-it-yourself kit, while eliminating the need for expensive test equipment and the services of an FCC-licensed technician.

Most of the FCC regulations apply to the transmitting section of a CB station, with little mention of the receiver, because of the possible interference to stations which operate on the same or adjacent channels. Off-frequency operation or distorted signals, which spill over into nearby channels, seriously affect the intelligibility of transmissions. Also, the power limitation of five watts keeps the range of a station on a local basis. Operators far removed from each other should be able to simultaneously transmit on the same channel. The technical restrictions on equipment ensure *clean signals*—an important factor in the crowded radio spectrum.

SPECIFICATIONS

Most CB sets are *transceivers* (Fig. 3-1)—a combination of transmitter and receiver. The transceiver design offers the twin advantages of compactness and economy. Duplication of circuits is avoided by making certain sections of the unit serve alternately in the send and in the receive positions. An audio amplifier, for example, boosts

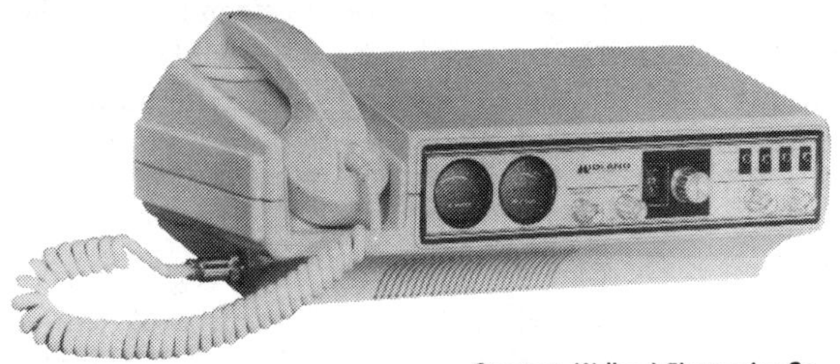

Courtesy Midland Electronics Co.

Fig. 3-1. A typical large base station transceiver.

the signals for the speaker when the set is on receive, and amplifies the microphone signals when the switch is thrown to send. The need for two power supplies is avoided in a similar manner. Some transceivers can double as a pa system, as shown in Fig. 3-2. Choosing an appropriate model is a matter of matching the features of the unit to the intended application.

Receivers

The least expensive of the CB transceivers utilize the superregenerative circuit and work well when the appli-

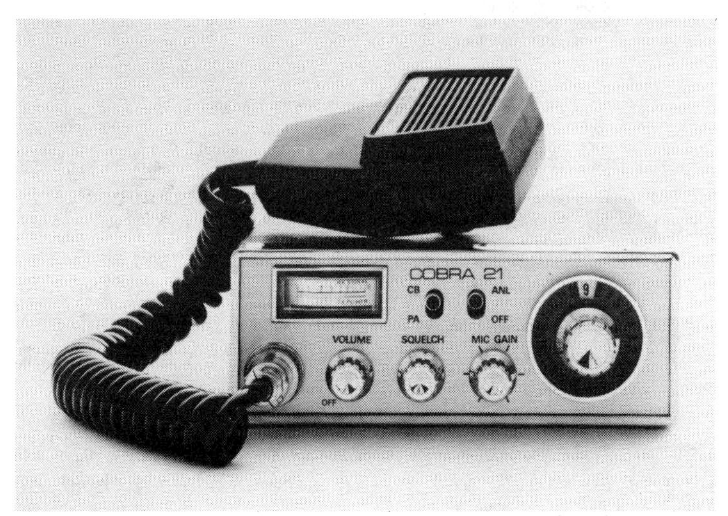

Courtesy Dynascan Corp.

Fig. 3-2. This small transceiver can double as a pa system.

cation is in an area where the Citizens band is not crowded. Its sensitivity, or ability to boost the signal and not the atmospheric noise, rivals that of the superheterodyne. Simplicity of design yields the long-term benefits of lower maintenance costs and less frequent component failure.

Selectivity—A disadvantage of superregenerative receivers is lack of selectivity, i.e., the abilty of the set to reject stations on nearby channels. Strong signals close

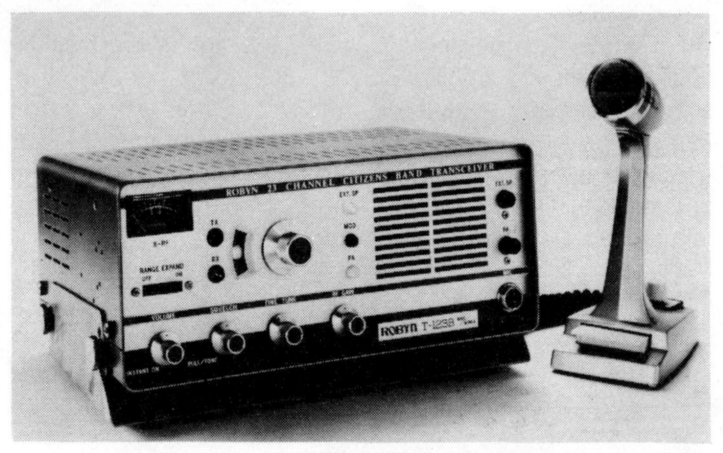

Courtesy Robyn International

Fig. 3-3. A typical superheterodyne transceiver.

to your operating frequency will enter the broad response pattern of the receiver and often block communications. Another limitation is that the circuit does not lend itself to crystal control. (This should not be confused with the transmitter, which is *always* crystal controlled.) Finally, the superregenerative type produces a background hiss in the speaker which could prove annoying when the band is monitored for long periods of time.

The superheterodyne transceiver (Fig. 3-3) is not afflicted with selectivity problems to the extent of the superregenerative, and it can accommodate a host of special features. Evaluating a *superheterodyne* is mostly a process of counting and comparing specifications. Since each feature will add to the overall cost, each one should be judged in terms of the intended use of the station.

Is the receiver equipped with a radio-frequency (rf) amplifier? This state contributes much to sensitivity and selectivity. Selectivity normally increases with the number of receiver stages. Although stages rarely exceed one in number, the intermediate-frequency (i-f) section may contain two or more.

More complex sets employ a system known as *dual conversion,* where several i-f stages sharpen the *response* of the receiver, making it immune to a type of superheterodyne fault known as images. These are false signals that can interfere with your channel, although they are produced by stations on other frequencies (not necessarily in the Citizens band). When incorporated into a particular model, this circuit refinement is usually well-advertised by the manufacturer.

Noise Limiting—Mobile transceivers should have some form of noise-limiting device since pulses of energy generated by automobile or boat ignition systems can seriously interfere with reception. There are several measures for eliminating noise at the source—from spark plugs, alternator, and voltage regulator—but they are ineffective against noise emanating from vehicles other than your own. The measure of a good noise limiter is how effectively it reduces noise without producing excessive distortion or reduction of the audio.

Squelch—Another valuable feature in the receiver of CB sets is *squelch.* This circuit will mute the speaker until a signal is received. A squelch is especially convenient when the receiver is left on for long periods of time in anticipation of a message. The constant background noise and assorted static crashes in the speaker can disrupt the quiet atmosphere of an office or prove distracting while driving. With the squelch switched on, the speaker comes to life only through the action of a received signal. The simple squelch system is not *selective;* other stations operating on the channel can also disable it and become audible. The *selective call* system, where the squelch is overcome only by your remote stations, is available.

Transceiver Tuning

One of the important questions in the choice of CB equipment deals with multichannel operations. In multichannel units a front-panel switch permits the choice of several transmitting frequencies. The immediate advantage is that communications can be shifted to a clear channel in the event of interference. Another benefit is that transmissions may be addressed to one group of remote units while excluding others (for example, trucks on one channel, passenger cars on another). Multichannel operation is useful where channels are set aside for specific purposes: marine, distress, civil defense, etc. Only traffic pertaining to these classifications is conducted on the predesignated channels.

The system used for receiver tuning is closely coupled with the multichannel operation of the transmitter. The simpler units have *continuous* tuning over the complete Citizens band. If the transmitter is of a single-channel design, the receiver tuning dial is left on the same channel with only occasional need for retuning. But, where multichannel operation is used, the continuously tuned

Courtesy Lafayette Radio and Electronics

Fig. 3-4. This CB transceiver has an extra, tunable band for monitoring vhf stations.

receiver can present a problem. The operator must tune his receiver to other channels with precision or the transmission will be missed. The tuning error is worsened by the dial calibration employed on most CB sets: channel numbers which appear on the panel are only rough indications of frequency. Drifting due to heating and mechanical "play" often causes the receiver to be off frequency.

Most transceivers utilize a crystal-controlled receiver. The operator manipulates a channel selector which clicks precisely into place, and the receiver is electronically locked on the channel by receiving crystals. The transmit channel, ganged to the same switch, simultaneously shifts into place. This system has been traditionally used in

Courtesy Pearce-Simpson, Inc.

Fig. 3-5. A small mobile transceiver on top of power supply for 117-volt ac operation.

commercial communications equipment because continuous tuning is slow and inaccurate. Sometimes a CB transceiver has a second band (Fig. 3-4) for monitoring very high frequency (vhf) stations in the 150- to 174-MHz range.

Compared with receivers, transmitter sections offer few special features except for multichannel operation.

The input power to the final rf stage of the transmitter must not exceed five watts and the crystal frequency should have a tolerance of ±.005%. Single sideband, a special form of transmission, is discussed in Chapter 5.

Power Supply

After a transceiver is selected on the basis of features versus cost and application, the major consideration is the power supply. Virtually all CB radios have provision for two types of primary power sources (Fig 3-5) : 117-volt ac and 12-volt dc. There are sets which offer only one of these power supplies. If the transceiver has interchangeable power supplies, select the one which matches the power source—automobile or boat storage battery, or the 117-volt ac house current. The trend, however, is toward the combined supply: usually 12-volt dc and 117-volt ac. A small unit for operation in the field, shown in Fig. 3-6, comes complete with rechargeable power pack.

Courtesy E. F. Johnson Co.

Fig. 3-6. This small unit comes complete with rechargeable power pack.

Versatility is attained with the dual supply, since the set may be quickly switched from mobile to fixed station operation. The changeover is usually effected by choice of the proper power cable.

CB RADIO KITS

The choice between buying a kit or a unit assembled by a manufacturer is a highly individual one. With reasonably good workmanship, kit-built equipment should match the performance of a factory model of comparable circuitry. If you do not calculate your assembly time in terms of dollars, the kit can yield the sense of satisfaction which most people associate with the do-it-yourself project. After the unit has been in service, some of the repairs and maintenance can be performed by the kit builder with knowledge acquired during assembly. According to FCC regulations, however, there are sections which may only be serviced by a licensed technician—primarily the frequency determining circuits which are prewired at the factory.

4

How the Transceiver Works

The basic design of the transceiver, whether it be superregenerative or superheterodyne, varies in the number of operating features. Thus it is possible to explain the operation of transceivers with two fundamental block diagrams. They serve as a kind of road map to trace the route of the signal as it is processed. While the transceiver is in the receive mode, the signal path is from antenna to speaker. In the send, or transmit mode, the signal originates in the microphone and ultimately radiates from the antenna.

TYPICAL SUPERREGENERATIVE SET

Referring to Fig. 4-1, note that the circuit is divided into four major sections—receiver, audio, transmitter, and power supply—each enclosed by dotted lines. In order to direct the signal to the proper section, switch/contacts (marked SW-A through SW-D) are shown at various points throughout the diagram. This usually represents a relay with four sections. When the mike button is pressed, the switch contacts changeover simultaneously and the signals will be routed correctly. The power-supply and audio sections then serve alternately for the receiver and transmitter. The transceiver cannot transmit and receive at the same time as a consequence of this. Each party must wait until the other transmission is completed before he can respond.

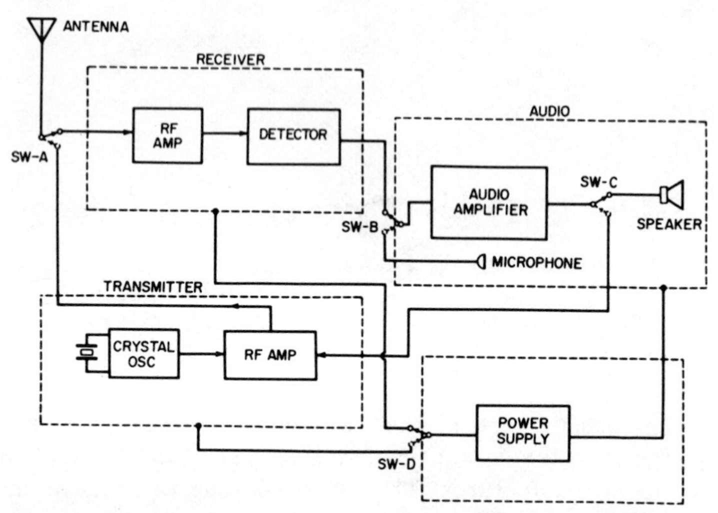

Fig. 4-1. Block diagram of a typical superregenerative transceiver.

Receive Mode

Fig. 4-1 shows the transceiver in its receive position. The antenna, at the upper left of the drawing, is placed to intercept the radio waves originating from the transmitting station. Signals of all frequencies travel down the antenna and are diverted to the receiver—the first major section—or more specifically, its rf amplifier. Signals reach this point through SW-A, the first section of the send-receive switch. If the switch were in its alternate position (transmit), the antenna would be connected to the transmitter.

RF Amplifier—The rf amplifier performs several functions. It amplifies or boosts the radio frequencies induced in the antenna; these voltages are extremely low in level and are measured in millionths of a volt. Also, the signals are tuned here. The customary tuning capacitor associated with rf amplifiers is usually eliminated in the superregenerative set. Rather than amplifying individual channels and rejecting all others, the rf amplifier is broadly tuned to select all 23 channels of the Citizens band. Isolation of the *detector* is another purpose of the rf amplifier. The detector, the second stage of the receiver, is susceptible to the swinging of the antenna in the wind. Unstable tuning would result if the rf amplifier did not act as a buffer against this effect. Another form of isolation is also provided: the detector oscillates and produces a signal of its own which could couple into the antenna and produce interference to nearby receivers. This radiation is curbed by placing the rf amplifier between the antenna and the detector.

Detector—The transmission system in CB is amplitude modulation (a-m). The voice is superimposed onto a radio wave which carries it between transmitter and receiver. The function of the detector is to remove the voice signals from the radio-frequency carrier and pass them on to the audio section for amplification. The detector also serves both as tuner and rf amplifier. Initally, the rf signals (with voice modulation) are applied to a tuning circuit in the detector. The control is a tuning dial on the front panel of the transceiver, which the operater rotates for the desired channel. Once selected, the signal is applied to the input of the stage, which builds it up by virtue of its amplifying action.

There is a limitation to the output of the detector which is dependent on how much signal is available at its input. To overcome this limitation, the *regenerative* detector is employed. By feeding a small portion of the amplified output signal of the detector back to its input, it is possible to boost the amplification. With this added input (identical to the original signal from the rf amplifier), the detector produces a greater output. But the radical increase in detector signal has a built-in limitation, which can be described in terms of a familiar example: place the microphone of a public-address system close to the speaker and a loud howling or whistling (called feedback) will be audible. This is produced by the microphone picking up a sound and feeding it to the amplifier. The sound, in turn, is fed back to the microphone from the speaker; consequently, the sound oscillates between microphone and speaker, becoming louder and louder as the feedback condition intensifies. The regenerative detector utilizes feedback for increasing its efficiency, but ultimately it suffers from the same feedback effect.

A circuit modification which overcomes the interfering howl in the detector is *superregeneration*. Here, the detector is permitted to go into oscillation to attain maximum amplification, but the howl is quenched. This action is possible because human hearing does not respond to tones much over 17,000 or 18,000 Hz per second. If the detector is forced to break in and out of oscillation at a rate which is too fast for the ear to perceive, the detector can operate with optimum feedback (consequently, maximum amplification) without any interfering howl. The quenching frequency often used in the superregenerative set is approximately 20,000 Hz.

Audio—The audio section of the receiver follows the detector. Note that SW-B, the second section of the send-receive switch, transfers detector output to the audio amplifier (not to the microphone). At this point in the circuit (the output of the detector) audio signals are very low in level and must pass through two stages of audio amplification before they are at sufficient power to drive the speaker to adequate listening levels.

Power Supply—The power supply, by SW-D, energizes the receiver when the set is in the receive mode. The audio portion of the circuit, however, is always powered, since it is used in both transmit and receive positions. Only the signal connections to the audio amplifier are switched.

Transmit Mode

During transmitter operation, the switch is in the send position, as shown by the dotted lines on the four switch sections. The crystal (at the lower left of Fig. 4-1), which determines the transmission channel of the transceiver, is a tiny piece of quartz, ground and lapped to extremely close tolerances and mounted in a plug-in container. For multichannel transceivers, more than one crystal is employed with an appropriate channel switch. The crystal, working in conjunction with a tube, generates electrical vibrations to precisely control the frequency of the transmitter. The *overtone* crystal was developed because a crystal cut specifically for a 27-MHz band is very fragile and therefore unable to pass sufficient current. If a crystal is cut to vibrate at a specified frequency, it will be found to produce overtones, or harmonics (vibrations, which are multiples of its fundamental frequency). Through special techniques, a crystal can be manufactured to favor its third harmonic—the most popular type in CB equipment. Therefore the fundamental of a 27-MHz crystal will be 9 MHz. To avoid confusion, manufacturers mark their crystals with the Citizens band or actual operating frequency. The crystal oscillator contains the frequency-determining elements of the transmitter, which should be adjusted or repaired only by a licensed technician.

RF Amplifier— Second in the transmitter line-up is the rf amplifier. Its purpose is to raise the low output of the crystal oscillator to the rated output power.

Modulation—The modulation for the transmitter originates at the microphone. The voice sets up trains of air movement which strike the diaphragm of the microphone. This mechanical action is converted into audio voltage and applied to the audio amplifier; as in the detector, the low-level audio is raised in strength. Since the send-receive switch is now in the transmit position, the audio goes to the transmitter rather than to the speaker. The modulation process of impressing the voice signal onto the rf carrier occurs at the output of the amplifier transmitter. As in the conventional a-m system, the strength (amplitude) of the rf carrier is varied according to the applied audio signal. When the audio amplifier section functions in this manner, it is called an a-m *modulator*.

Radiation—The modulated rf is passed to the antenna through the SW-A section of the send-receive switch, and transmission occurs as the carrier currents leave the antenna in the form of electromagnetic radiation. These

The reasons for altering the incoming frequency with the local oscillator are twofold: the selectivity of an amplifier is more favorable at lower frequencies, and the i-f amplifier can be designed to operate efficiently on a single frequency.

Assume that an i-f frequency of 1750 kHz (1.75 MHz) is arbitrarily selected. When the set is tuned to 27.065 MHz (Channel 9), the local oscillator will produce 25.315 MHz. When the two signals (local oscillator and received signal from the antenna) combine, or heterodyne in the mixer, the result is a difference of 1750 kHz. Other mixing products are present in the mixer output, but the i-f amplifier is fix-tuned and only accepts 1750 kHz. The voice modulation on the rf carrier remains unaffected by the mixing process.

If the tuning dial is removed to another channel, the local oscillator will similarly shift and generate a frequency which is always 1750 kHz below the channel frequency. The net result is that all incoming channels will be converted to the i-f frequency having a low numerical value to provide sharp selectivity. Also, the i-f stage operates best on its nominal 1750 kHz with no need for the compromise incident to tunable amplifiers. The ultimate in superheterodynes uses the double conversion system, in which the i-f frequency is converted to a lower value for a second time.

Crystal control of the local oscillator is a means for providing rapid, precise tuning of the receiver. In this modification, the variable tuning dial gives way to a switch and one or more crystals. When the switch is clicked onto a channel, the crystal generates the exact local-oscillator frequency for that channel. In some transceivers a selector switch gives the operator the choice of either continuous tuning or crystal-controlled reception, thus adding to the versatility of the set and equipping it for any mode of operation.

The output of the i-f amplifier is applied to the detector for removal of the voice modulation from the carrier wave (now on 1750 kHz). At this point in the circuit, the noise limiter functions to clip the sharp pulses of interference which often mar reception. The squelch performs its job by cutting off the operation of the first audio amplifier unless a signal is being received. Output of the audio amplifier then proceeds to the balance of the audio amplification section and speaker.

POWER SUPPLIES

The primary source of electricity determines the type of power supply used to energize the transceiver. For home use, the supply is usually of the transformer-operated type. Tracing the pathways shown in Fig. 4-3, the line cord is plugged into the 117-volt ac source. The transformer then provides two levels of output voltage for

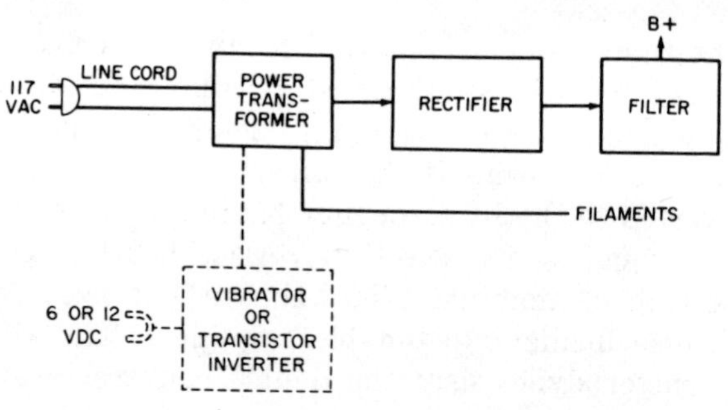

Fig. 4-3. Block diagram of a typical power supply.

tube sets: approximately 300 volts ac and 6.3 volts ac. The higher voltage proceeds to the rectifier where it is changed to pulsating dc, and on to the filter where it is smoothed out to pure dc. At the output, the pure dc is termed the B+ voltage and is used to power the plate element of the various tubes in the transceiver. The lower 6.3-volt output, which does not have to be rectified and filtered, heats the tube filaments, lights the pilot lamps, and energizes the relays (if any are used). In solid-state sets, house current is stepped down and filtered to 12 volts dc.

Where the primary power source for a tube set is a storage battery (automobile, boat, etc.) a vibrator or transistor inverter is added as shown in Fig. 4-3. Its purpose is to change the pure dc from the battery to interrupted dc, a type which the transformer can handle. The transistors in a solid-state set operate directly off the mobile power (12 volts dc) so inverters or vibrators are eliminated.

5

Single Sideband

Single sideband (ssb) is a specialized form of transmission found in CB manufacturers' most expensive models (Fig. 5-1). The circuits are more complicated than those of regular transceivers, which transmit a conventional a-m signal. Although a sideband transceiver can be used to communicate with any other transceiver, it performs best when operating in the sideband mode. A regular transceiver receiving sideband transmission will hear gibberish that sounds like a band of chattering monkeys. Speech is not only garbled but completely unintelligible, except to another sideband transceiver.

In exchange for this, the sideband CBer gains tremendous efficiency. A sideband signal of 5 watts becomes the equivalent of a 40-watt conventional signal. The audio is not 8 times louder at the receiving end, but you can expect anywhere from 1½ to 2 times the normal operating range. Many operators report that it is not the range improvement that is most important but filling in of former "holes" in the coverage of their station. This creates more usable channels within the existing band. Because of the far-reaching advantages of sideband many commercial, amateur, military, and marine stations oper-

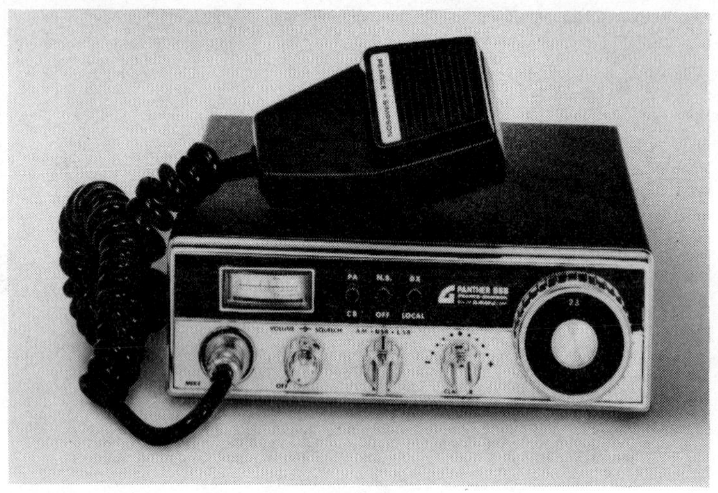

Courtesy Pearce-Simpson, Inc.

Fig. 5-1. A typical single-sideband transceiver.

ating in the high frequency (hf) band (3 to 30 MHz) have already switched to it.

Sideband is a very efficient form of conventional a-m. It may come as a surprise, but the carrier of a conventional a-m transmitter carries no audio intelligence. This is seen on the S-METER of the receiver; it shows a steady signal, generated by the carrier, that does not fluctuate with the voice. The useful audio signal actually lies in two sidebands on either *side* of the carrier. If you could tune a radio dial across an a-m signal, you would encounter a lower sideband, the carrier, and an upper sideband. The S-METER would kick in step with the two fluctuating sidebands but not for the steady carrier. Why produce a carrier if it does not bear useful information? It is unavoidable while generating sidebands inside a conventional transmitter. It is produced by a result of audio and radio signals mixing in the final amplifier during modulation.

Let's look at a conventional three-part signal to see how it develops into single sideband. As shown in Fig. 5-2, a person is speaking into the microphone (1000-Hz audio tone). The major output of the transmitter, in this in-

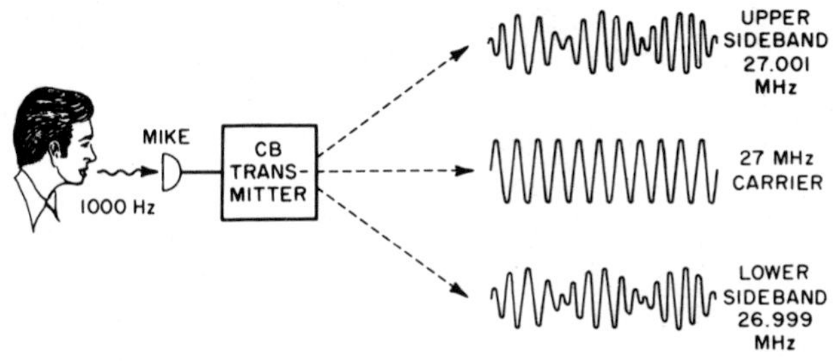

Fig. 5-2. Illustrates the carrier signal plus upper and lower sidebands.

stance, is the steady carrier on 27 MHz. An upper sideband forms exactly 1000 Hz above that carrier, on 27.001 MHz, while the lower sideband falls exactly 1000 Hz below the carrier on 26.999 MHz. If the person speaks in normal conversation, the sidebands would consist of all the audio tones spaced by their frequencies above and below the carrier.

Another way of viewing the conventional a-m signal is shown in Fig. 5-3. The sidebands are mirror images of each other, and one can be eliminated without losing any of the original audio tones in the speaker. The most critical aspect of the signal, however, is how the power is consumed. If the transmitter is putting out a total of 3 watts (a typical amount), the carrier takes fully ⅔ of it, or 2 watts (Fig. 5-4). At the same time, the remaining

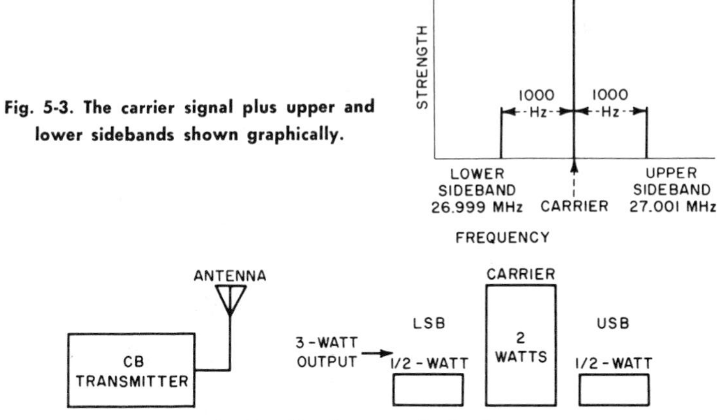

Fig. 5-3. The carrier signal plus upper and lower sidebands shown graphically.

Fig. 5-4. Signal power distribution.

power is split between the 2 sidebands, for ½ watt each. It would be more efficient to have all the available transmitter power in one sideband, since this is all that is needed to bear the intelligence.

This is what a single sideband transceiver does, as shown in the block diagram of Fig. 5-5. First, a carrier signal is developed by oscillator crystals in an early stage of the transmitter. This signal, plus that of the microphone, is fed into a balanced modulator where the audio and radio frequencies combine. The result of the mixture is the two sidebands—upper and lower—but no carrier signal. The carrier signal is introduced into this stage in a fashion which permits it to create sidebands and cancel itself out at the output.

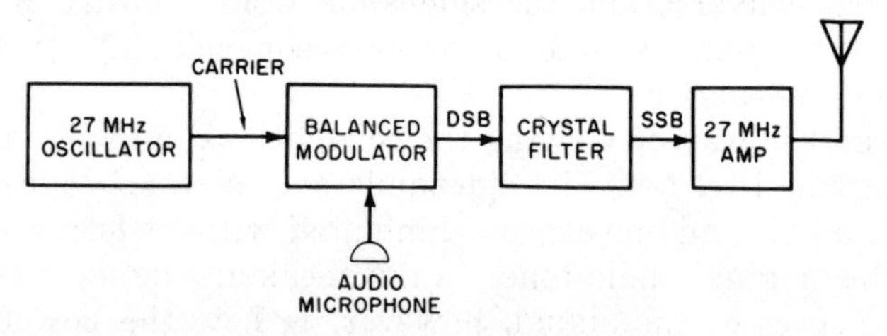

Fig. 5-5. Simplified block diagram of single sideband transmitter.

Recall that only one sideband is needed to carry all the necessary audio information. This accounts for the next stage, a crystal filter, which is sharply tuned to only one sideband. It admits only the desired sideband while suppressing the other. The result now is a true single sideband signal which is fed to the 27-MHz amplifier. In a conventional final amplifier, most of the available power is wasted on a steady carrier, but in a sideband final amplifier, all the rf power is concentrated on a single sideband where it does the most good.

PEAK ENVELOPE POWER (PEP)

*P*eak *e*nvelope *p*ower is used to rate the output of a sideband transmitter. In a conventional transmitter, rf power is measured by observing the steady carrier signal for a typical output of 3½ watts. This technique cannot be used to rate sideband transmitters because there is no carrier output. Only when you speak into the microphone is there a transmitted signal. It is more convenient, therefore, to measure power when the sideband hits its peak, or maximum, value while an audio signal is applied. This yields peak envelope power of up to 12 watts; this is much higher than for conventional transmitters but still within the regulations, which rate power in average watts.

SIDEBAND RECEIVERS

A sideband receiver has several different stages and controls compared to those of a regular a-m receiver. Inside its circuits are narrow band filters which reduce atmospheric noise pickup. These filters are much narrower than on normal receivers because an incoming sideband signal takes up less space than a double-sideband signal. For the same reason, the sideband receiver can receive twice the number of channels—46 instead of 23. For example, an operator may choose Channel 12, but then select its upper or lower sideband for transmission.

Sideband receivers have an internal *b*eat-*f*requency *o*scillator (bfo) which makes code audible. As mentioned earlier, the carrier helped create the sideband signal and was then suppressed in the transmitter. The missing carrier is needed again to reverse the process—converting the sideband back into audio. The receiver does it much more efficiently with a low-power bfo than the regular CB receiver does with the original carrier.

Typical Sideband Controls

The sideband transceiver has several controls that differ from those of regular transceivers (Fig. 5-6), and their position and function will be discussed in the following paragraphs.

RF Gain—In a regular CB receiver, an automatic gain control senses the strength of the carrier and protects the receiver against overload (heard as a garble or distortion). Since sideband provides no carrier to operate a

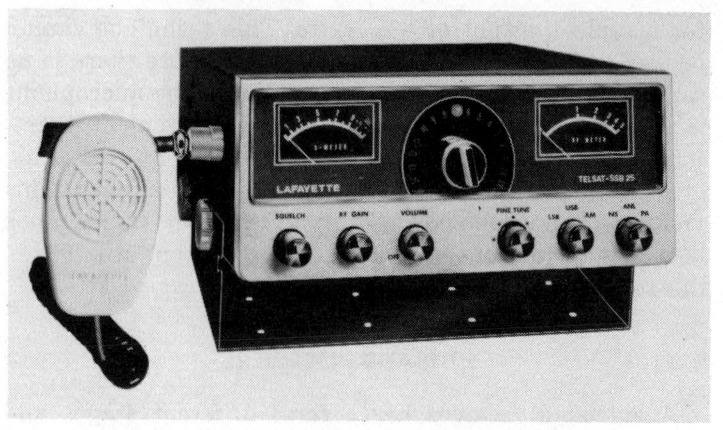

Courtesy Lafayette Radio and Electronics

Fig. 5-6. A single-sideband transceiver with additional controls.

conventional automatic gain control circuit, the RF GAIN control is provided so you can manually adjust the sensitivity of the receiver. It is usually used to reduce the strength of powerful sideband signals.

Clarifier (*Fine Tune*)—This ia a *fine-tuning* control that clears up the incoming voice. This is sometimes called VOICE LOCK as shown in Fig. 5-7. As mentioned previ-

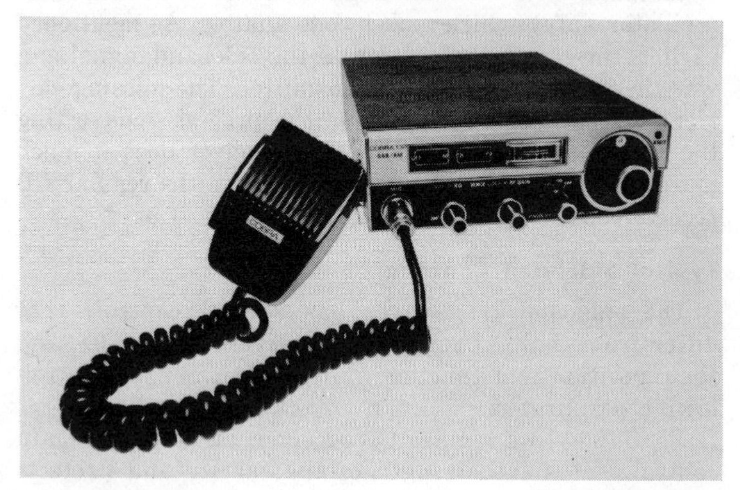

Courtesy Dynascan Corp.

Fig. 5-7. A single-sideband transceiver with Voice Lock.

ously, the receiver supplies the missing carrier (through its bfo) to recreate the original audio signal. The carrier must be inserted with tremendous frequency accuracy—within about 50 Hz of the original carrier. To avoid unreasonably expensive circuits, the operator makes the final frequency adjustment with the CLARIFIER control. It takes some practice, but the knack of clarifying usually takes this form: if the voice sounds extremely low or guttural in pitch, the control is turned until the tones grow higher. When the voice sounds like Donald Duck, on the other hand, try to reduce the pitch. At a precise point the voice is heard with excellent clarity and strength.

AM/LSB/USB—With this selector you can choose one of three modes for each CB channel. When it is placed on AM, the transceiver transmits and receives as would any conventional transceiver. Thus it is compatible throughout the CB system and allows communication with any other operator. At this time the efficiency of the signal drops to that of any transceiver.

Your choice of LSB or USB depends on prior arrangement with other operators or common practice in the local area. Sometimes you may switch from one sideband to another to avoid interference from other stations.

Since the selector switch has three positions (AM/LSB/USB) that can be used on each of the 23 channels, some manufacturers use it as a basis for a claim that their transceivers operate on "69 channels."

NB—This control is a noise blanker which is a more elaborate form of noise limiter.

The remaining controls and features of a single sideband transceiver, like SQUELCH and S-METER, are similar to those of conventional equipment. Most sideband transceivers, however, provide all 23 channels and have a multipurpose meter on the front panel for measuring S-units, rf power and, in some cases, the condition of the antenna system. If you have never heard a sideband signal, monitor Channel 16—the unofficial channel where sideband stations often operate. On a conventional receiver, however, the lack of a bfo and rf gain causes the signal to be completely unintelligible.

6

Antennas

A CB antenna system assumes enormous importance—it is the primary, single element for determining range. The 5-watt limitation on transceivers leaves little margin for power loss in the antenna installation. From the profusion of antennas on the market, it is possible to choose a model well suited to any particular application. A "multielement stacked array" can make a 5-watt transceiver sound like 80 watts to a distant station—but is useless in certain modes of operation.

FCC regulations on antennas mostly concern height; the antenna must not extend more than 20 feet over a man-made or natural structure. If the antenna is nondirectional (the most popular type), it may mount on a tower, mast, or pole as long as it doesn't exceed 60 feet over *ground* level.

RANGE

As radio waves are emitted from an antenna, they are comparable to an inflating balloon; they expand outward in all directions at once, becoming less dense with distance. If no obstacles are interposed between transmitting and receiving points, the range can be phenomenal. This was dramatized in 1960 when an American space probe transmitted data from a point 22 million miles from earth with its tiny 5-watt transmitter. Of course, the techniques used in space-probe reception are quite sophisticated, but the comparison points up the potential of an unimpeded radio wave.

Height

The basic rule for achieving maximum operating range states: mount the antenna as high as possible. This affords the radiation an opportunity to clear nearby trees and buildings. In mobile applications height is impractical since the antenna is mounted on a vehicle. But, if the fixed-station antenna is high, and in the clear, the height deficiency of the mobile antenna can be offset to a significant degree; the fixed station will transmit stronger signals and receive weaker ones.

Antenna height will also overcome the horizon. The ability of a radio wave to follow the curvature of the earth depends on its frequency. At low frequencies the wave hugs the earth's surface, thus following the curve over the horizon. This is ground-wave transmission, which yields reliable coverage for standard broadcast stations and other low-frequency services. But CB occupies the lower edge of the band. Ground-wave propagation in this sector of the radio spectrum is considerably reduced and gives way to the direct wave. The characteristics of direct-wave transmission are somewhat similar to light waves. Due to its inability to follow the earth's curvature, the direct wave shoots off into space, as illustrated in Fig. 6-1.

Note in the illustration how the direct wave travels from the fixed to the mobile station. The reflective properties of the hill indicate the importance of antenna height. Part of the wave from the fixed station clears the hill, while the lower portion is reflected upward.

The 27-MHz band of CB is subject to a third type of propagation—*the sky wave.* This is produced by the action of the ionosphere. When radio waves strike this elec-

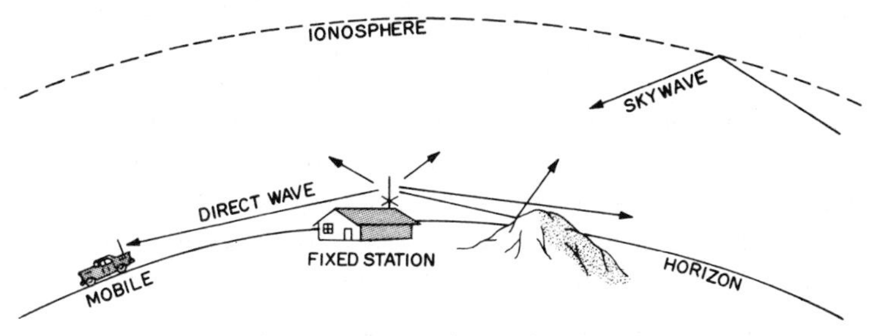

Fig. 6-1. Behavior of CB radio waves.

trical "mirror" high in the earth's atmosphere, they bend and return earthward. In an effect known as skip, waves can cover tremendous distances as they hop between the earth and the ionosphere.

Communication via sky waves is illegal for CB stations. By no means shall two CB stations communicate with each other over a distance of more than 150 miles. Sky wave is described here since it can be a source of interference. When conditions are favorable, stations located anywhere from several hundred to several thousand miles away will be heard as if they were originating locally. Fortunately, sky-wave propagation lasts for just a few hours per day, and only during certain periods of the 11-year sunspot cycle.

Other Variables

The range of a particular CB system cannot be predicted on a mileage chart. The variables of antenna type, height, and location and the terrain make this impossible. Certain general categories, however, can be stated. The shortest range is encountered in heavily populated urban areas, especially between two automobiles. Under rapidly changing conditions, results are erratic. Tests in a large city reveal that fairly consistent communication is possible within a radius of about 1 to 3 miles. This distance increases when one of the transmitting points is elevated above ground level. The fixed station, for example, with an antenna located on top of a building, could double the mobile range. In one test, where an antenna was mounted on the roof of a 22-story building (over 200 feet high), the talking distance to a mobile station increased to 10 miles.

Open terrain, with few hills, aids the transmission path. A 10-mile average is normal under these conditions. Where signals are eclipsed by the shadow of a hill, it is often feasible to drive to the top of the hill and gain the necessary height. Range over open water is excellent—

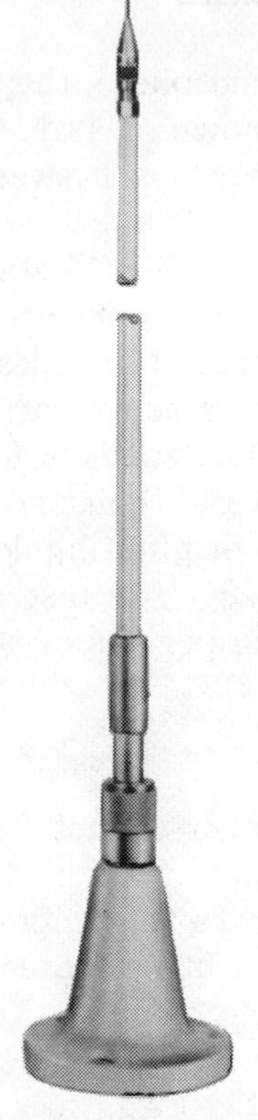

Courtesy The Antenna Specialists Co.

Fig. 6-2. Antenna made for small boats.

the average is 20 miles. Salt water, favorable to the ground-wave component, imparts three times the signal distance of dry land. See Fig. 6-2 for a typical antenna that requires no ground for use on small boats.

Maximum range in CB is rarely over 20 miles. This is attained through a combination of height and the high-gain antenna. As detailed under the section devoted to directional beams, this distance can be achieved under ideal conditions which primarily exist only between two fixed stations.

HALF-WAVE DIPOLE

A discussion of specific antenna types should start with the half-wave dipole (Fig. 6-3) ; it is simple in design and serves as the standard against which virtually all other antennas are measured. *Half-wave* refers to the physical length of the radio wave. Citizens Radio is assigned to the 11-meter band, which reveals that one wavelength is approximately 35 feet long (after converting meters to feet). In actual practice the antenna is cut to a half-wave, or somewhat over 17 feet.

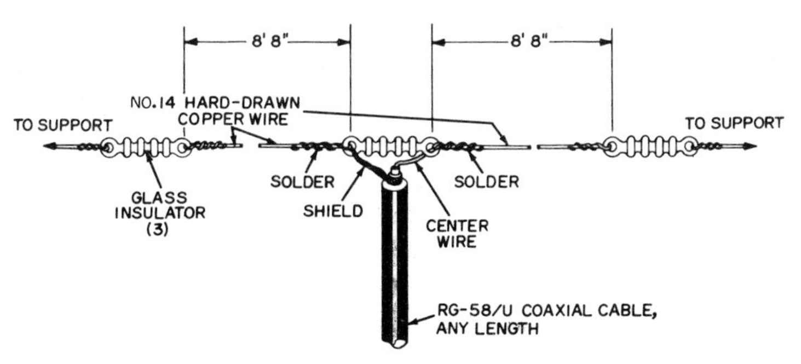

Fig. 6-3. Construction detail of a dipole antenna.

The half-wave dimension fulfills two important requirements: it permits the antenna to absorb maximum power from the transmitter, and it is a good compromise between physical size and efficiency.

The signal output of the transmitter is not fixed; it surges in and out of the antenna, changing direction at the rate of 27 million times per second (its assigned frequency). If its path in and out of the antenna were not of the correct physical length, the wave would buck itself and reduce overall power by cancellation. The half-wave is one of the basic configurations which peaks up, or resonates, at the applied rf.

A consideration in the use of a dipole, or any antenna, is its radiation pattern. The dipole is basically bidirectional—maximum energy occurs broadside to the wire. If the antenna runs north-south, most of the energy will radiate in an east-west direction.

Polarization

As seen in Fig. 6-3, the dipole is mounted horizontally. This influences the antenna property known as polarization. If the dipole were mounted vertically, the radiated waves would change from horizontal to vertical polarization, and the result would be a shift in the radiation pattern to nondirectional. Instead of a "figure 8" distribution of energy, the pattern would look more like an expanding doughnut reaching out in every direction at once. Best efficiency takes place when both transmitting and receiving stations have antennas with similar polarization. Since the antennas on vehicles are almost always vertically mounted, the trend in CB is to use vertical polarization for the fixed station. Cross polarization occurs where two stations use antennas which lie in different planes; the result produces a reduction in signal strength.

BASE STATION ANTENNA

The ground plane (Fig. 6-4) is an example of a quarter-wave antenna (sometimes used for base station operation). Its most significant feature is that its length is one half that of the dipole. Its principal dimension is somewhat over 8 feet, which provides ease of mounting for both mobile and fixed station operation. As a vertically polarized antenna, the ground plane may be supported by a single mast.

The ground-plane aspect of the antenna provides the missing half of the radiating element when compared to the half-wave dipole. Ideally, the vertical element should sit just above the surface of the earth supported by an insulator as is done in standard broadcast stations. One or more towers, in excess of 100 feet high, comprise quar-

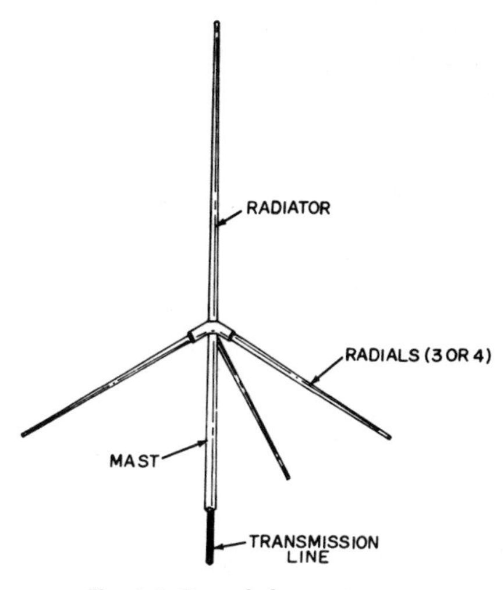

Fig. 6-4. Ground-plane antenna.

ter-wave radiators which function in close proximity with earth ground. But CB on 27 MHz utilizes direct rather than ground-wave propagation, and consequently requires height. The ground plane is a means of simulating ground at a point above the earth. Physically it appears as a group of radial elements which extend outward from the quarter-wave length vertical rod. Commercial versions of the ground plane have three or four elements which may mount anywhere from a right angle to a drooping 45° angle, relative to the vertical rod. The radials make the ground-plane antenna physically independent of earth ground, while providing the necessary electrical effect.

The ground plane was once a popular antenna, but improved models have replaced it because they radiate stronger signals. An example is the half-wave vertical (Fig. 6-5). Instead of 9 feet in height, it extends to 17 feet 10 inches. Near the base of the antenna is a matching transformer to couple the bottom of the antenna to the coaxial cable with the proper electrical match.

Another model is the ⅝ wave antenna (Fig. 6-6) which extends to just under a maximum antenna height of 20 feet. It also gives rise to one of the most popular base

Courtesy Cush Craft Corp.

Fig. 6-5. Half-wave vertical antenna.

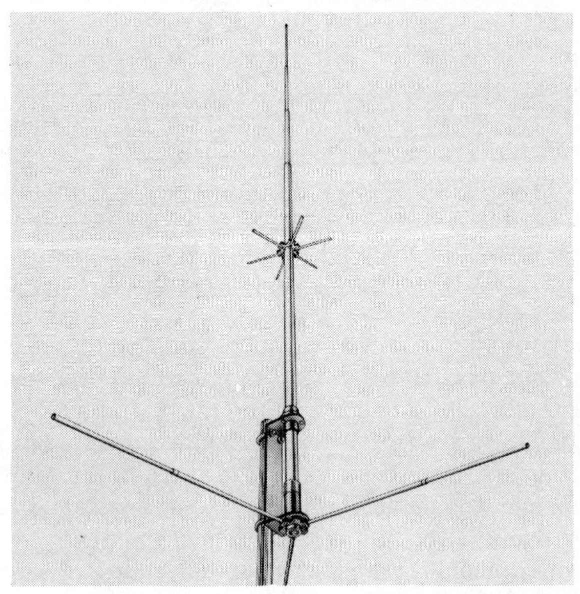

Courtesy New-Tronics Corp.

Fig. 6-6. A colinear base station antenna.

antennas, the "colinear." The term is derived from antenna elements which are stacked in the vertical direction.

MOBILE ANTENNAS

The most popular mobile antenna, the whip shown in Fig. 6-7, is actually a ground-plane variation. There is, however, no need for radial elements since the car body itself presents a large mass of metal just below the vertical rod.

The radiation pattern of the whip is essentially nondirectional. But, in actual installations, the pattern is subject to departures from a perfect circle. This stems from the fact that the distribution of metal below the whip is not equal in all directions. Various obstructions, too, will alter the pattern. The bumper-mounted whip, for example, is partially blocked by the trunk deck and rear fender of the car.

LOADING COILS

Virtually all antennas may be physically shortened by a loading coil—inserted in series with the elements. The effect is to electrically lengthen an antenna which is physically cut to less than a quarter wave. This has given rise to a number of clip-on types which may be conveniently attached to the rain gutter of the car (just above the side windows). The standard ground plane and whip are also available in shortened versions.

The loaded antenna will always have less range than the full quarter-wave whip. Where short range is adequate, however, the small size of these units is used to advantage for limited space applications and mounting versatility.

The continuously loaded whip combines the loading coil and radiating element in a single conductor. It is constructed of a spiral wire which extends from the bottom

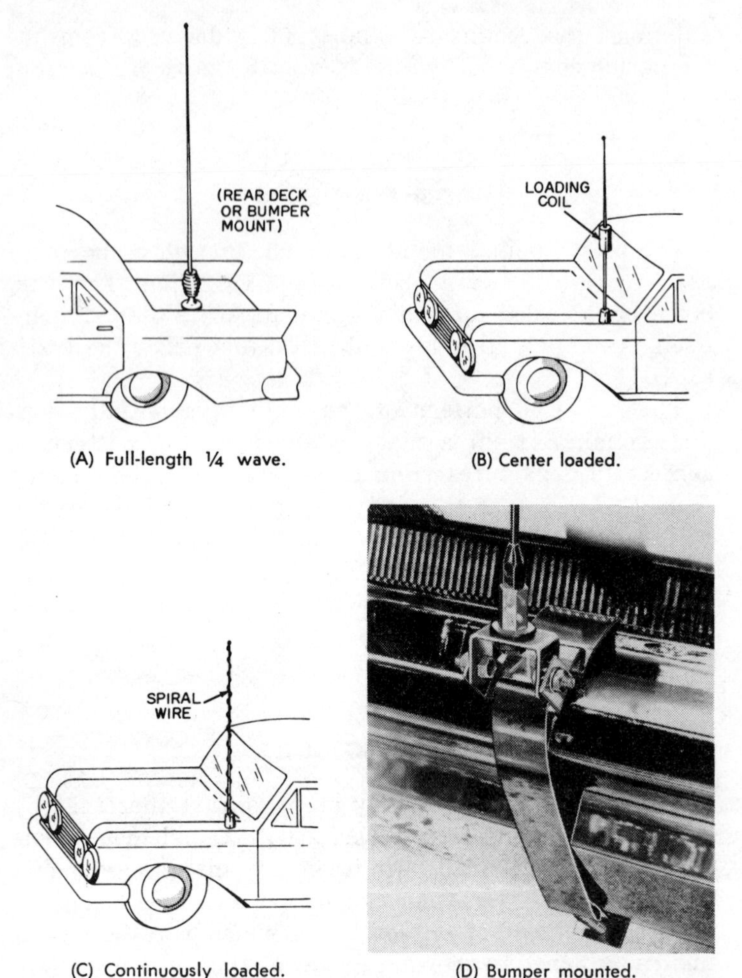

(A) Full-length ¼ wave.

(B) Center loaded.

(C) Continuously loaded.

(D) Bumper mounted.

Courtesy Hy Gain Electronics Corp.

Fig. 6-7. Basic mobile whips.

to the top of the whip rod and is usually encased in plastic for support and flexibility.

DIRECTIONAL BEAM

Among the antennas most efficient for communication between two fixed points is the directional beam (Fig.

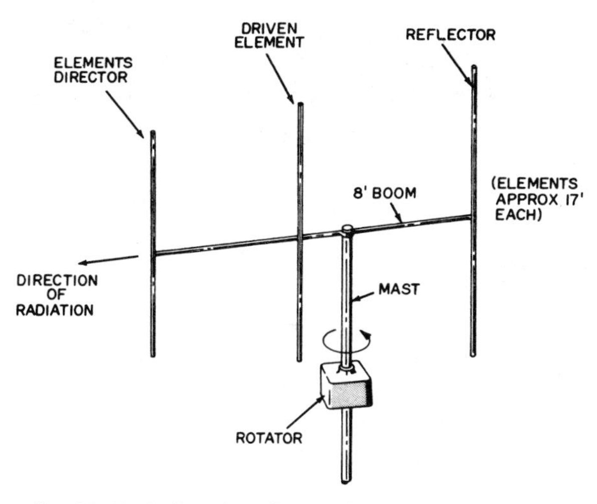

Fig. 6-8. Vertically polarized three-element directional beam.

6-8). With its array of elements, most of the radiated energy can be concentrated into a narrow arc and aimed at the desired station. Multiplication of power occurs since most of the energy which would be wasted in undesired directions is channeled into a narrow beam.

The heart of the beam antenna is the half-wave dipole. Assuming that it is positioned in a north-south direction, it will radiate the conventional "figure 8" pattern with lobes of energy propagated to the east and west. As the various elements are added, this pattern is modified. If a horizontal rod of proper length is placed several feet to the east of the dipole, it will act as a reflector and bounce the radio wave back toward the dipole. The result is a near doubling of power in the westerly direction. The pattern of this simple two-element beam is greatly sharpened by the addition of one or more directors on the opposite side of the dipole. These rods pick up energy from the dipole and reradiate it toward the west. Whereas several directors will narrow and reinforce the beam, only one reflector is used to return the arc of electromagnetic energy to the desired direction.

Stacking is a technique to further increase the power gain of a beam (Fig. 6-9). Two director-dipole-reflector combinations can be vertically stacked and tied together by a suitable matching device. The limit to the amount of stacking is mainly a physical one. A complex array soon becomes unmanageable and imposes a heavy load on the mounting mast and boom.

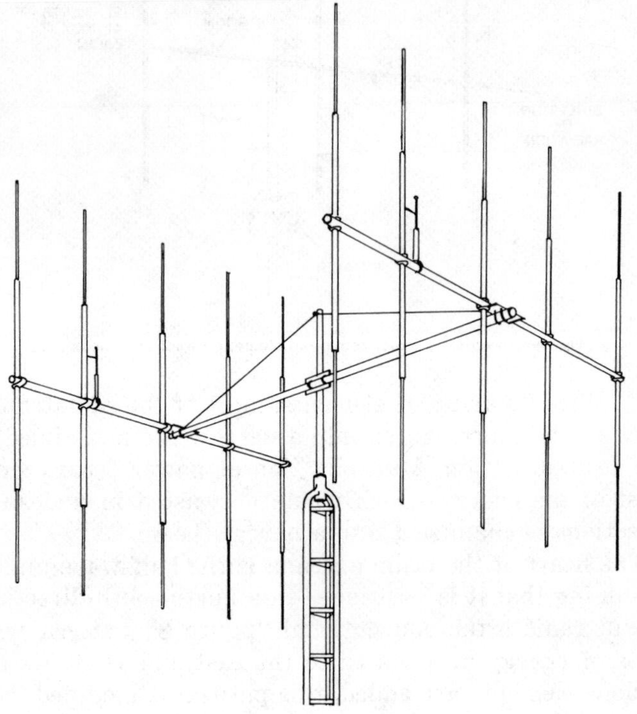

Courtesy Mosley Electronics, Inc.

Fig. 6-9. Stacked five-element beam.

The fact that a 5-watt signal can be reinforced is not a violation of the FCC law. The regulations limit the input and output power of the final radio-frequency stage of the transmitter. As long as the antenna does not exceed 20 feet over a natural or man-made structure (which excludes towers, masts, and poles), its ability to impart signal gain is acceptable.

The directivity of the beam antenna renders it a poor choice for certain applications. When a fixed station is in contact with several mobiles, the vehicles will not always be in the signal path of the beam. It is possible to orient the beam with a rotator, but this is a marginal procedure. An example of efficient use of a beam is between home and office, where the distance may not easily be covered by a ground plane antenna. Two beams oriented toward each other can supply maximum range and reliability of communication. Some specialized beams, like the one in Fig. 6-10, can be switched to either vertical or horizontal polarization to reduce certain types of interference.

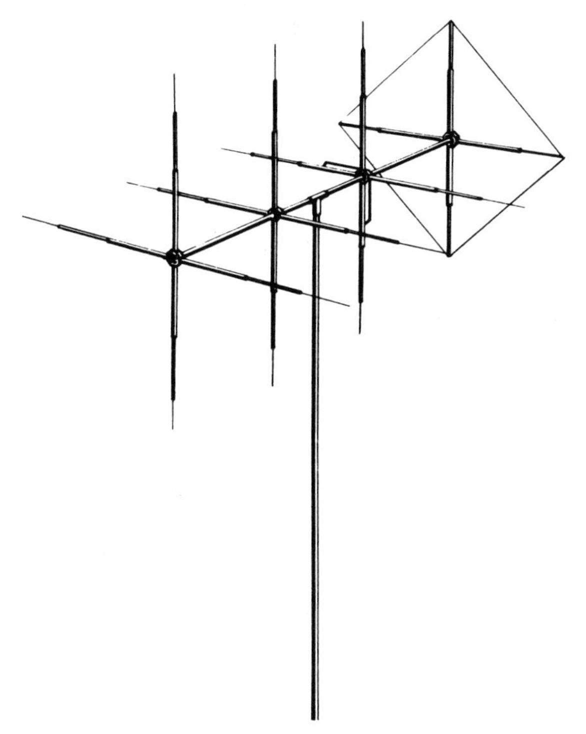

Courtesy Avanti R & D, Inc.

Fig. 6-10. Directional beam for either horizontal or vertical polarization.

TRANSMISSION LINES

Unless the antenna simply plugs into the transceiver (for very short-range work) a transmission line is employed to carry radio-frequency power between transceiver and antenna. Only a cable which meets precise specifications will effectively conduct with minimum loss. The coaxial cable has been almost universally adopted for CB equipment since it does not radiate along its length. It does, however, place the power in the antenna where it counts.

At the antenna jack of a typical transceiver the impedance is approximately 52 ohms. This value indicates that a transmission line whose impedance is also 52 ohms can be connected to it. It follows that the other end of the line hooks into a 52-ohm antenna. Through observing and matching impedances, maximum transfer of energy will take place between these elements. Most of the antennas for CB are pretuned at the factory and designed for 52-ohm input. An ideal coaxial cable is RG 58/U.

7

Accessories

There are a dozen or more instruments, attachments, testers, and other devices in the CB accessory category. Some are useful for tune-up and troubleshooting; others are handy during installation and checkout. All are simple to connect and operate. They also improve operating convenience or indicate the need for repair by a technician. This chapter starts with two or three of the most popular CB accessories, then discusses more exotic devices that appear on dealer shelves or in the electronic catalogs.

FIELD-STRENGTH METER

Although the field-strength meter has little more than a meter, a coil, and a small telescoping antenna, it indicates much about the transmitter signal. It is a miniature receiver that senses the signal from the antenna and indicates it on a relative scale. Since the meter is operated by the air signal, it will indicate the condition of every important element in the transmitting chain—transmitter, cable, and antenna. By keeping records, you will know what the relative reading is when everything is known to be operating correctly, and the figures serve as a future reference to confirm trouble.

Field-strength meters are available in different models. One model has a magnetic base that can be placed on a car dashboard for monitoring the transmitter output each time the mike button is pressed. Since the air signal can drive the meter directly, this model requires no battery. Other models have batteries and built-in amplifiers to improve their sensitivity and pickup range. An earphone jack may be provided so the quality of the transmitted audio can be monitored.

Courtesy Radio Shack

Fig. 7-1. A field-strength and swr meter combined in one instrument.

Since the field-strength meter is so useful, it is often incorporated as part of another instrument (Fig. 7-1). Before you acquire a field-strength meter, decide whether you want a separate device or one that is teamed with a more complex instrument. The advantages of a separate field-strength meter include small size, portability, and lower price.

SWR METER

The swr meter is installed in the coaxial cable and indicates valuable information about the transmitter and antenna. Swr, or *s*tanding *w*ave *r*atio, is the amount of transmitter power applied to the antenna compared to the amount reflected. The reflected part represents loss of power due to mismatch and should be held to the lowest possible value. The swr of an efficient CB system should be below 2 to 1. Factors which increase swr include dam-

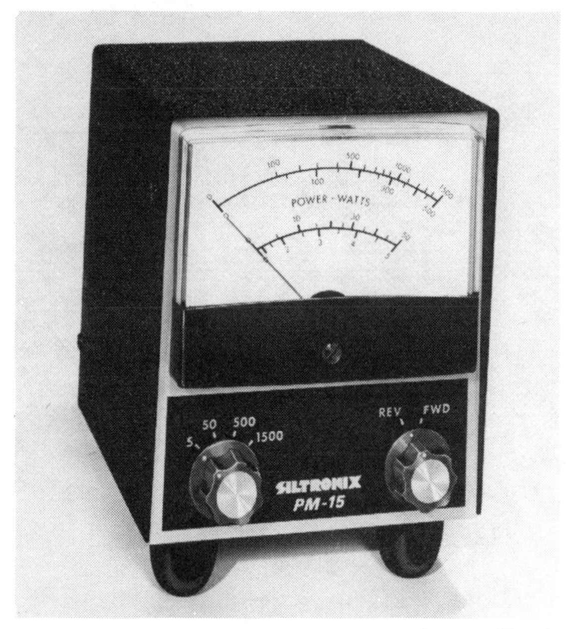

Courtesy Siltronix

Fig. 7-2. A typical wattmeter.

aged antenna elements, poor transmitter tuning, shorted or open coaxial cable, and mounting an antenna too close to a nearby mass of metal.

An swr indicator (also called "bridge" by some manufacturers) is an excellent companion to a field-strength meter. Since swr is tuned for lowest reading while field strength is observed for a maximum, tune-up errors can be avoided. When a system operates at best efficiency, both indications are seen simultaneously.

WATTMETER

This instrument, also placed in line with the coaxial lead, overcomes a disadvantage of the field-strength meter. The wattmeter gives an absolute, not relative, indication of power; that is, you can read power output of a

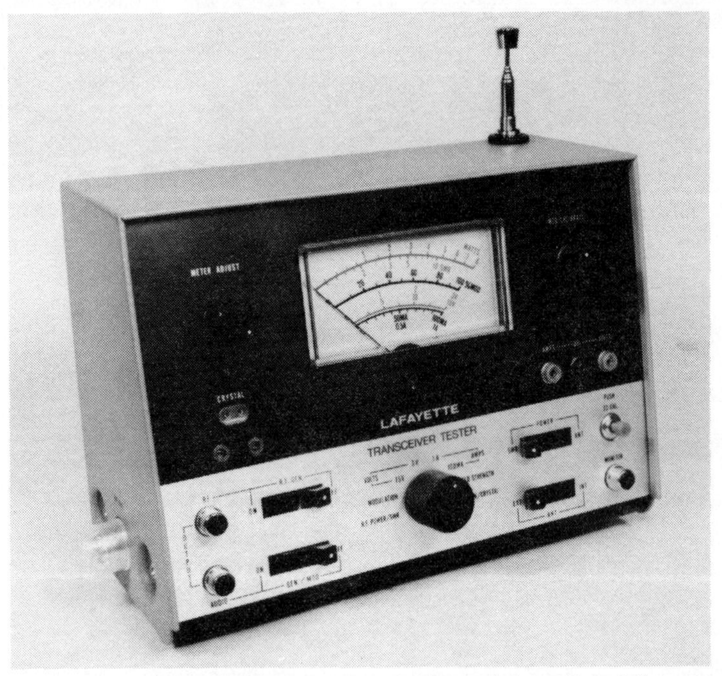

Courtesy Lafayette Radio & Electronics

Fig. 7-3. A typical transceiver multitester.

transmitter directly in watts (Fig. 7-2). This avoids a condition where you read satisfactory swr in an antenna system, but the transmitter is producing low output. The wattmeter indicates whether the transmitter is generating rated power, which often falls between 3 and 4 watts. The instrument is usually designed to be left in the line without consuming output power.

COMBINATION TESTERS

These are a combination of several instruments mounted within a single enclosure (Fig. 7-3) and called something like "CB Tester" or "CB Transceiver Multitester." They have a selector switch to switch in various functions, like WATTS, SWR, FIELD STRENGTH, and PERCENTAGE MODULATION through a single meter. More sophisticated models have provisions to test CB crystals, voltage, and amperes (for checking a mobile power source), and can also provide a 27-MHz test signal for receiver troubleshooting.

DUMMY LOAD

As mentioned in Chapter 9, you can use a No. 47 pilot lamp for transmitter test purposes. It is a dummy load

Fig. 7-4. Dummy load matches transmitter output.

which converts output power into heat to prevent interference on the air as the set is being adjusted. A more elegant device is a commercially available dummy load (Fig. 7-4). Its advantages are better shielding to contain the signal and a special noninductive resistance which maintains an excellent match to the transmitter output (52 ohms).

LIGHTNING ARRESTERS

If the coaxial cable is properly grounded (the shield goes to a good electrical ground), this affords some protection against lightning. It is better to use the commercially built lightning arrester (Fig. 7-5) that is designed to mount in line with the coaxial cable. It has a built-in

Fig. 7-5. Lightning arrester used with coaxial cable.

spark gap to provide a good path to ground for an electrical discharge. There is, however, a ground terminal on the arrester and it must be connected to a good electrical ground to be effective.

ANTENNA SWITCHERS AND MATCHERS

Many coaxial switches (Fig. 7-6) are available for making quick changes in cable connections. Let's say you have two antennas—a nondirectional standard type and a directional beam. You can instantly choose between them with a two-position coaxial switch. You can also use

Courtesy Hy Gain Electronics Corp.

Fig. 7-6. Two- and three-position coaxial switches.

a two-position switch in reverse to connect two transceivers to one antenna; then choose the set that will feed the antenna. There are larger models which have up to six different positions available.

Antenna matchers are small boxes (Fig. 7-7) with tuned circuits that help match the coaxial line to the antenna. Usually they are not necessary because the line and antenna are both rated at approximately 50 ohms and should connect directly together with no significant loss of power. Sometimes, however, mounting an antenna on

Courtesy E. F. Johnson Co.

Fig. 7-7. Antenna matcher matches coaxial line to antenna.

a car causes the impedance of the antenna to drop to some low value and create a mismatch to the line. The matcher unit, which installs in the line near the antenna, can correct it. When using a meter to install a matcher, adjust the matcher for the lowest reading.

MICROPHONES

CB manufacturers supply microphones as standard equipment with their sets, but special mikes are popular items. They usually have internal amplifiers or circuits which keep modulation percentage high (Fig. 7-8). This helps the operating range, but without "splatter" interference which is caused by overmodulation. Another advantage is that soft and loud voices are smoothed out and kept at a high modulating value. Wiring diagrams for connecting such microphones to the CB set are usually supplied by the manufacturer.

Courtesy Turner Microphone Co.

Fig. 7-8. Special CB microphone.

PHONE PATCH

These instruments are widely used in amateur radio to provide low-cost telephone calls for servicemen overseas wishing to talk to their families back home. Cost is reduced because most of the distance is toll-free; the message is carried overseas by radio through amateur stations. The phone patch connects the radio equipment into the telephone line. A relaxation in telephone company rules about such attachments in the late 1960s stimulated the introduction of several models for the CBer.

The advantage of a phone patch in CB is not the same as that of amateur radio. Since radio transmission distances are always short, there is no important toll-saving by substituting a CB set for the phone line. The benefit is that two parties may speak to each other directly for better communication without having the CB operator repeat their words. For example, during a disaster or emergency, a public-service official in the field can speak into a mobile CB microphone, be picked up by a distant CB station, and have his voice "patched" to a telephone at his home office.

Fig. 7-9 shows in greater detail how the phone patch works. Assume someone wishes to place a call through

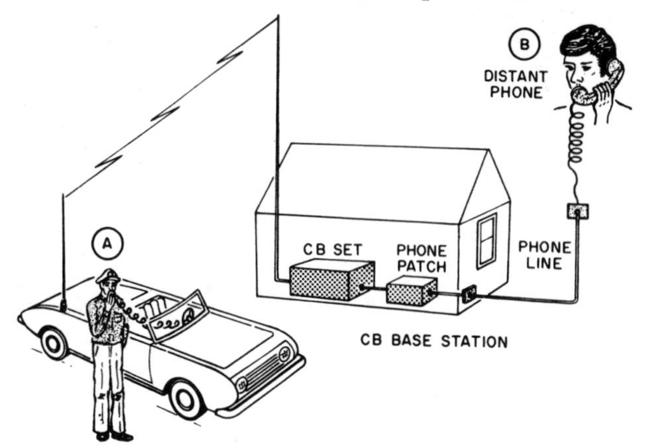

Fig. 7-9. Illustrates the operation of a CB phone patch.

your mobile CB set. You contact another CB operator known to have a phone patch installed at his base station and you request that he dial the desired telephone number. Once the party called is on the line, the two parties can converse with each other. Only one talks at a time, as in normal CB operation. Note how the calling party speaks into your microphone, reaches the CB base station, and connects into the phone line. The distant party answers through the CB base station and is heard in your loudspeaker.

Installation of a phone patch, only at the base station, requires three basic connections. Wires go to the CB loud-

8

Installation

A carefully planned installation pays big dividends in operating ease and efficiency. Mounting CB equipment demands little in the way of special skills or tools—rarely more than a screwdriver, pliers, hand drill, and other inexpensive implements are required. The transceiver is typically a package with main components housed in a single cabinet. This arrangement lends itself to rapid installation and can conveniently be used on a semipermanent basis, i.e., occasionally changed from mobile to fixed-station service.

FIXED STATION

The simplest installation is a fixed station which requires less than a mile range. This is often encountered in farm applications and certain industrial uses where the units serve as a "wireless intercom" between nearby vehicles and buildings. Most transceivers will accommodate a short whip mounted directly on the cabinet. The unit can be placed on a desk, table, or shelf, and simply plugged into a wall receptacle. No special power lines are required—the power drain of the transceiver averages less than that of a tv set or hi-fi system. Place the unit where air circulation is abundant, being especially sure to avoid blocking the ventilation holes.

Location

Within the limits of a mile, the transceiver should operate satisfactorily inside a wooden frame building on the first-floor level. The radio energy will penetrate wood, plaster, and other nonmetallic materials with little attenuation. Metal surfaces, however, are opaque to radio waves and act as effective reflectors. If the building is not of "bird cage" construction, a considerable amount of energy will find its way out. Location of the CB transceiver in a steel constructed building often results in blind spots in the radiation pattern as signals reflect in an unpredictable manner. The antenna should be placed at least several feet from any large metallic object, which includes window screens, filing cabinets, etc. A further improvement is possible with a window-mounted antenna—its level of performance is somewhere between the indoor and roof antenna. There is a possibility, however, that signal strength will be reduced if the antenna is located on the opposite side of the building from the receiving station.

Installation of an antenna in an attic should not be overlooked. This location yields the multiple benefits of simple installation, protection from the elements, and ease of maintenance. In its sheltered location, the life of the attic-mounted antenna is practically indefinite. The height advantage over the whip fastened to the transceiver cabinet makes it a good choice for moderate range applications—assuming, of course, that the roofing material is nonmetallic.

Outdoor Antennas

The high outdoor antenna will, in most cases, be essential at the fixed station. As the control unit of the network, it should be able to contact the most distant mobile unit in the service area. The coaxial cable which carries power to the antenna influences the location of the transceiver in the room; the cable should be as short as possible. Thought should be given to the routing of the cable as it runs from the window to the roof; the shorter the span of cable, the lower the attendant power drop. The method of feeding the cable to the outside will depend on the individual installation. Where feasible, drill a hole in the base of the window frame and feed the cable through. The hole should be caulked to prevent the entry of moisture. An alternative is to chisel out a channel in the window sill to a depth which permits the window to fully close without striking the cable. Unlike the twinlead used for tv transmission line, the coaxial cable does not suffer losses when positioned close to a metal surface. But the tv standoff is useful to support the cable run from the window to the roof. It is made with a small insulated hole to accommodate the cable and comes with either a screw-base (for fastening to wood) or nail-base (for hammering into concrete or the space between bricks).

Much of the hardware for installation of tv antennas will serve admirably for CB use. Antenna mast sections are sold in 5- and 10-foot lengths. Brackets are available in wall, chimney, pipe, eave, and roof-peak mounts. Where masts rise higher than 10 feet, guy wire should be used. The rotators for tv antennas are usable with a directional beam of the light, three-element type.

The first step in selecting an antenna site on a roof is locating the spot which offers best mechanical support and maximum height. Although a chimney often fills the height requirement, it should not be used if large volumes of smoke and soot continuously issue forth. These may form conductive deposits and lower the efficiency of the insulated antenna elements. If a fixed directional beam is to be installed, the heading of the other station should be considered. Choose the mast location which best enables the signal to clear large obstacles.

When faced with the dilemma of height versus transmission line length, height should win out in most cases. In a 100-foot run of RG 58/U coaxial cable, close to half the transmitter power is lost in the cable. Thus, if another 10 or 20 feet of line allows the antenna to be positioned at a higher point, it should be mounted there without question. The obstacles which the signal would encounter at a lower position often cause more power attenuation than a moderate increase in the length of the transmission line.

One of the troublesome points in an antenna system is where the transmission line connects to the antenna terminals. An effective means is the use of a coaxial connector which provides the line with a friction and screw fastening to the antenna. This forms a strong mechanical joint. Many commercial antennas are equipped with this feature. But where screw terminals are provided for fastening the shield and center conductor of the cable to the antenna, extreme care should be used to prevent poor contact. Both conductors of the cable, shield and center, should terminate in lugs. There are lugs which provide mechanical support by crimping to the conductor insulation and effect a good electrical connection after being soldered. Once the lugs are installed on the ends of the cable, they are slipped under the screws and washers of the antenna terminals and tightened. All strain is removed from this point by running the cable through a standoff clipped to the mast. An added measure to preserve the efficiency of the antenna is to wrap the terminals completely with black plastic tape. This reduces corrosion and the leakage effects of material which may deposit between the terminals.

Sensible precautions can protect you against the major hazards of antenna installation—falling from a height and electric shock. Unfortunately, a good antenna site is often located near power lines. While manipulating the elements of an antenna, it is possible to brush against these lines and come in direct electrical contact with dangerous high voltages. It is wise to be extremely wary and perform as much antenna assembly on the ground as possible.

MOBILE

The power requirements of a CB transceiver are comparable to a standard auto radio. This eliminates many of the special considerations needed with higher-power sets. The conventional ignition system in modern automobiles is adequate without special generators, alternators, or high amperage batteries. Installation problems center on transceiver and antenna location and interference from ignition noise.

Mounting

The favorite spot for a transceiver in automobiles is under the dashboard, approximately a foot to the right of the steering column. This permits access to all operating controls, yet presents little interference to driver and passengers. While determining the precise mounting position of the transceiver, hold it in place by hand. It is not a good idea to locate it directly in the path of the hot air stream from a duct of the automobile heater.

Most manufacturers provide mobile mounting brackets for their transceivers (Fig. 8-1). Make a visual check under the dashboard for predrilled holes which match those of the brackets. Some automobile makers provide knockout holes under the dash. If these do not match the bracket holes, new ones must be drilled. The metal here is usually thin enough to permit the job to be done by a hand drill. Before applying the drill, be certain that no damage will be done to wires that run behind the dashboard; push them aside while drilling.

If brackets are not supplied for the transceiver, they may be made from angle iron, L-brackets (sold in hardware stores), or a sheet metal strap. Fasten them to the sides of the transceiver cabinet with sheet metal screws or nuts and bolts, being careful not to disturb any of the components inside the unit. Wing nuts or thumb screws are convenient where the transceiver is frequently removed from its mount.

If, for some reason, underdash mounting is not desirable, a second choice is under the front seat. Some automobiles have sufficient room here for a complete transceiver, but several disadvantages must be considered. Sound from the speaker is reduced, which affects intelligibility, especially when the automobile is traveling at high speeds. The operating controls are not readily accessible and a transceiver which has a push-to-talk button on the microphone must be used. Also, if the ON-OFF switch is not accessible, another must be installed in the power line.

Power Lines

After the transceiver is in place, two lines from the power plug are connected. Some power supplies will work

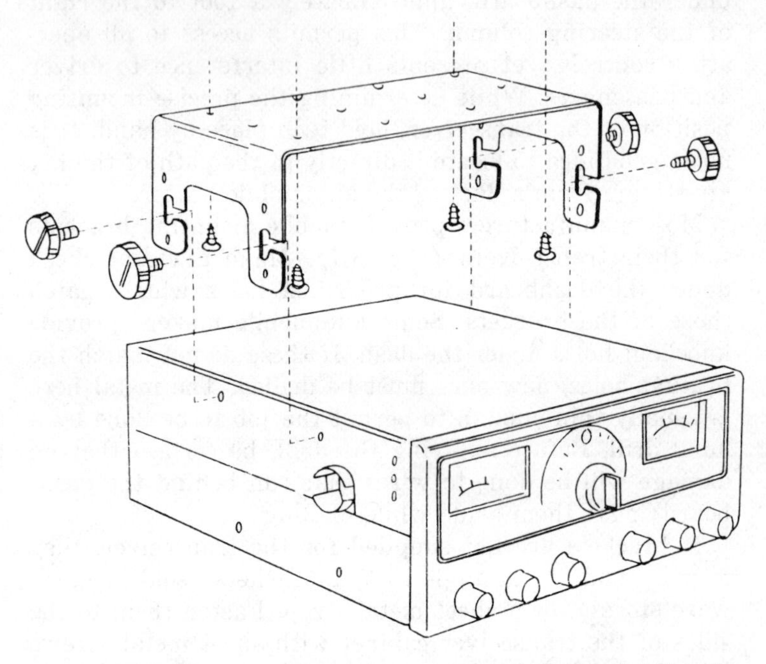

Courtesy Lafayette Radio & Electronics

Fig. 8-1. Typical mobile installation showing bracket mounted on topside of transceiver.

equally well with positively or negatively grounded automobile batteries, but this should be checked in the instruction manual. It can also be checked by looking for the plus and minus marking stamped on the battery terminals. The ground lead is usually a heavy braided wire connected directly to the chassis or frame. The ungrounded lead goes to a terminal protruding out of a small automobile enclosure which houses the starter solenoid.

The ground lead (Fig. 8-2) from the transceiver, often color-coded black, is fastened to the frame of the car as close to the set as possible. (If no ground lead is supplied, make one from braided wire.) Actually, the set should

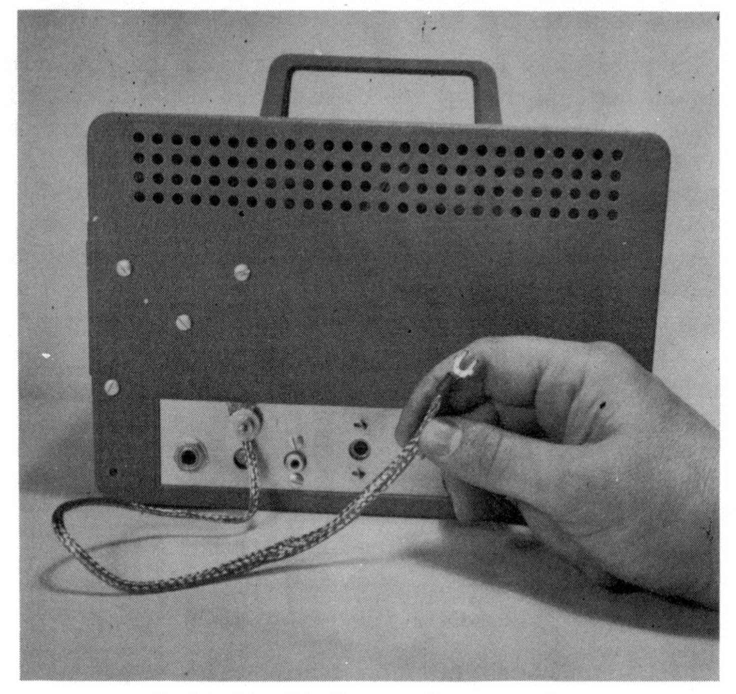

Fig. 8-2. Ground lead connected to a transceiver.

ground through its mounting brackets and antenna cable, but the ground wire will ensure a dependable connection and reduce the possibility of intermittent noise pickup. To establish a low-resistance electrical path, be sure to scrape away any paint or corrosion where the ground wire connects to the frame of the automobile.

The *hot* lead, usually red, can be slipped under the same terminal used for the automobile radio. This is often the accessory terminal behind the ignition lock and automatically cuts the power when the ignition key is removed.

In semipermanent installations, a plug is inserted into the cigarette lighter receptacle for power pickup. The cigarette lighter in many automobiles continues to supply power even though the ignition key is removed; so special care should be taken to avoid leaving the transceiver on accidentally, or the battery will run down in a few hours.

Whip Antenna

A good place for a whip antenna is on the rear left trunk deck. Mounting holes are drilled and the coaxial cable routed underneath the floor mats from the trunk to the transceiver. Standard lengths of coaxial cable are available for this use, but one can be made from RG 58/U with connectors to match the transceiver socket and the antenna base. If necessary, an antenna clip may be fastened to the appropriate point on the rain gutter of the automobile to hold down the whip when it is not in use.

Another whip location is in the space provided for the original automobile radio antenna. If you wish to avoid drilling holes, choose an antenna which mounts in the crack in the trunk lid. There are loaded whips designed for CB which substitute for the broadcast whip. These units will serve for both the transceiver and the automobile radio. Particularly suited for this location is the continuously loaded whip, extending approximately four feet.

NOISE SUPPRESSION

The automobile ignition system is a prolific generator of radio noise on the Citizens band. This interference can hamper communications considerably. The major source of noise originates in high-voltage impulses which fire the spark plugs. The alternator and voltage regulator also contribute to this hash. The most effective countermeasure is the noise limiter, a feature found in many superheterodyne transceivers. But noise suppression at the source will compensate for the deficiencies in the noise limiter, the reduction of speaker volume, and a decrease in intelligibility of the received signal.

Tracking down noise should be a step-by-step process, since the interference may arise from a single cause or a combination of effects. Before proceeding with the tests outlined here, check the ground connection of the transceiver. With the automobile engine running and the transceiver turned on, determine the effects of shifting the ground wire from one point of the car frame to another. Secure it to the ground area which produces the quietest reception, but try to keep the wire short. Also, check for any noise suppression measures which the automobile manufacturer has already installed.

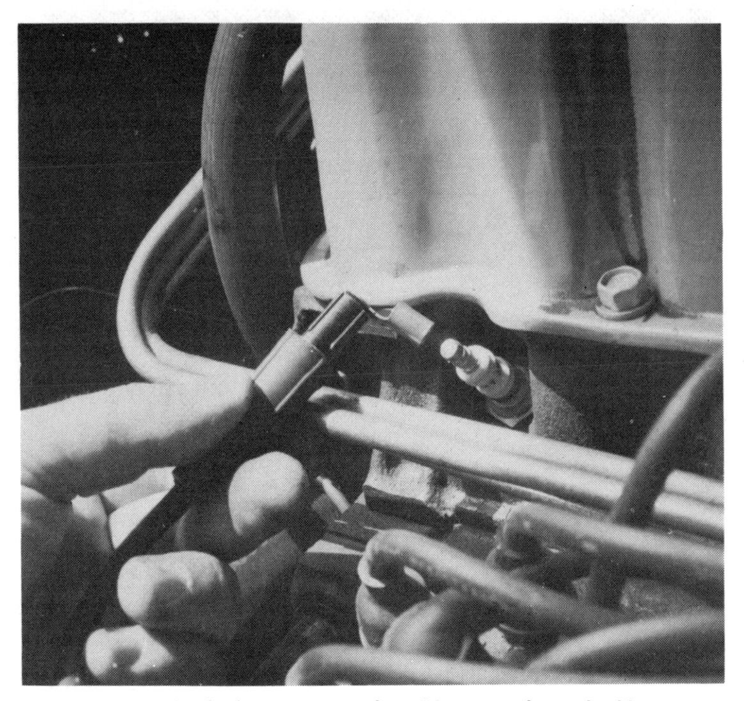

Fig. 8-3. Spark plug suppressor clipped between plug and cable.

These tests are listed in order of importance, ranging from the most common noise sources to the rarest ones. Switch on the transceiver, set its channel selector somewhere in the middle of the band, and idle the car engine. It is not necessary to receive a station during these steps. Turn up the volume control so the effects of the noise can be easily discerned. Try to ignore the sound of normal background hiss.

Distributor

During the first check, quickly depress the gas pedal. Listen for a train of sharp "popping" sounds which increase in speed as the engine revs up. This is produced by the spark plugs or distributor and can be reduced by suppressors (Fig. 8-3). These devices, available in automobile supply stores, simply snap into place. The cable to

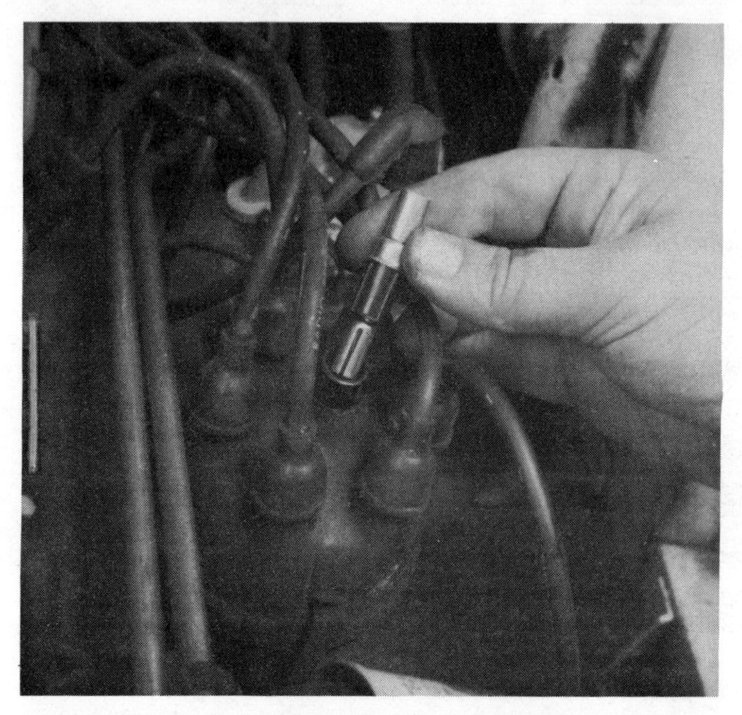

Fig. 8-4. Installation of distributor suppressor.

each plug is removed and the suppressor pushed on to the top of the plug. Each cable is then connected to the top of its suppressor. For the distributor, a single suppressor (Fig. 8-4) is plugged into the center hole after the cable is removed. If your automobile is due for new spark plugs, the resistor type may be installed—the suppressor is built into the body of the plug.

Alternator

The second test determines the noise effects of the alternator. Race the engine and listen for a howl or whining sound (almost musical) which lowers in pitch as the engine slows. Noise is produced by internal sparking and may be cured by installing an alternator filter and having the alternator slip rings checked and cleaned every 10,000 miles.

Fig. 8-5. Static collector springs installed on front wheel.

Wheels

A less common noise source is the front wheel axle. As the wheels rotate, a static charge is built up between the axle and wheel bearing, since they are insulated by a thin film of grease. The bouncing of the wheel causes contact between the two surfaces, and the static charge is intermittently dissipated. This might produce noise when the automobile is driven over approximately 20 miles per hour. Static collector springs (Fig. 8-5) positioned under the axle dust cap counter this action by establishing a constant electrical path between the axle and the bearing. These devices are often stocked by the automobile dealer for your make of automobile.

A static charge, similar to axle noise, can be generated by the inner tube and tire. The cure is the injection of antistatic powder (Fig. 8-6) into each of the tube valves.

Fig. 8-6. Special tool and packets of antistatic powder.

A small kit, complete with special tool and packets of powder, is available through electronic dealers.

Driving at various speeds over a bumpy road reveals poor grounds between certain areas of the automobile. The noise is intermittently keyed to the jarring of the road bumps. The technique of bonding is used to minimize this by providing low-resistance grounds between the motor block and the automobile chassis; the firewall and the motor block; and the muffler-tailpipe assembly and the automobile frame. Run a short length of heavy grounding strap between these parts, and be sure to keep the connections clean.

9

Tuning Up

The low power of CB necessitates careful tune-up; the receiver section should operate at peak sensitivity, and the transmitter must couple maximum energy into the antenna. Many of the tuning procedures can be done with very simple equipment, while others may be performed only by an FCC-licensed technician. The various adjustment screws and slugs used in the equipment are notoriously fragile—they are easily broken when tuned with an ordinary screwdriver. Damage can be avoided by using the alignment tool (Fig. 9-1) designed expressly for the job (available at electronic distributors). Many of these tools are made of nonmetallic material to prevent undue effects on the circuit while alignment is being done.

RECEIVER

Adjustment of the receiver requires no commercial license. Specific alignment procedures are usually covered in the manual which accompanies the kit- or factory-assembled sets.

Simple Alignment

The simplest alignment utilizes an off-the-air signal which can be provided by another transceiver in your vicinity. Avoid locating the sending unit too close to the set being aligned. This could lead to tune-up on false responses, which occur if one or more stages in the receiver

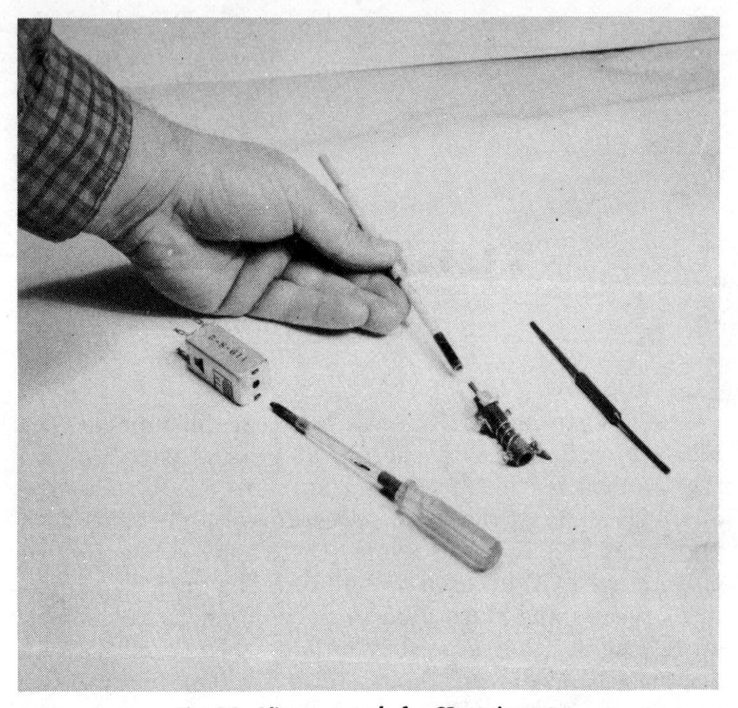

Fig. 9-1. Alignment tools for CB equipment.

section overload—a ¼-mile or more separation should prevent this. Listen for two distinct sounds in the speaker during the alignment process: the voice and the noise behind it. The most accurate indication that a particular stage is being peaked is the reduction in the amount of noise, rather than an increase in voice volume. Avoid adjusting any trimmer or transformer slug more than a few turns. These components are prealigned at the factory and only need "touching up" to compensate for the slight variations between transceivers. Excessive rotation of these adjustments usually indicates that something else is at fault, such as a wiring error or damaged component.

Certain adjustments will be more critical than others. Where tuning is extremely broad, try to set the adjustment to the center of its sensitivity range. Reducing the amount of received signal is helpful for perceiving the best setting.

Kits

If the wiring of a kit is not carefully done, especially around the i-f transformers, there is a possibility that some amplifier stages will go into oscillation during alignment. This produces a "mushy" or distorted quality to the sound in the speaker—often sharply reducing intelligibility. The remedy, after rechecking the wiring, is to back off slightly on the adjustment until normal reception is restored. Be sure that oscillation does not occur over any part of the receiver tuning range. In the superregenerative set there is a control which purposely sets up an oscillation during normal operation. This is the regeneration—turned up until a hiss is heard. Intelligibility is not affected since the hiss is reduced when a signal is received. Do not advance the control more than approximately one-eighth of a turn past the point where the hiss is heard or the set sensitivity will drop. When the control is properly set, the hiss should be present as the tuning dial is rotated from one end of its range to the other end with no incoming signal being received.

Instrument Alignment

Maximum receiver sensitivity is attained by alignment with a signal generator and a voltmeter. If this test equipment is available, the step-by-step instructions in the manual can be followed. The major precaution is to feed as little signal as possible into the receiver during the tune-up to prevent overloads. The basic procedure is to apply a 27-MHz signal into the receiver and note the dc voltage produced by the automatic volume control (avc) circuit. Starting at the last tunable stage in the set (the transformer before the detector), each adjustment is peaked for maximum reading on the meter. Any stage which breaks into oscillation will cause the needle to deflect rapidly up the scale.

TRANSMITTER

The transmit section of the transceiver has fewer tune-up stages than the receiver. But here is where FCC law must be observed. The reason for this is that improper alignment can result in off-frequency operation, interference to other stations and radio services, and excessive distortion on your signals. Actually, a "clean" signal is to your advantage; it often goes hand in hand with maximum power output.

For tune-up purposes, the transmitter can be considered in two major sections: frequency-determining circuit and output. The first section may be adjusted only by a first- or second-class radio telephone technician. These circuit elements include the crystal and the associated components which comprise the crystal oscillator. In consideration for the popularity of the kit transceiver, the FCC has waived the commercial license requirement—*if* the kit manufacturer provides a crystal oscillator which meets certain specifications. Most notable is that the crystal oscillator must be supplied from the factory as a pre-assembled, pretuned unit. Any adjustments which would cause improper operation cannot be made without breaking the seal of this subassembly. The manufacturer must certify that the kit is designed and furnished in accordance with applicable FCC specifications.

Dummy Load

The final checkout on a completed kit should be done with a dummy load (Fig. 9-2). This consists of a No. 47 pilot lamp soldered to an appropriate connector and plugged into the antenna output jack on the set. Tuning can be done with the dummy load without radiating signals; output power is consumed in the form of heat and

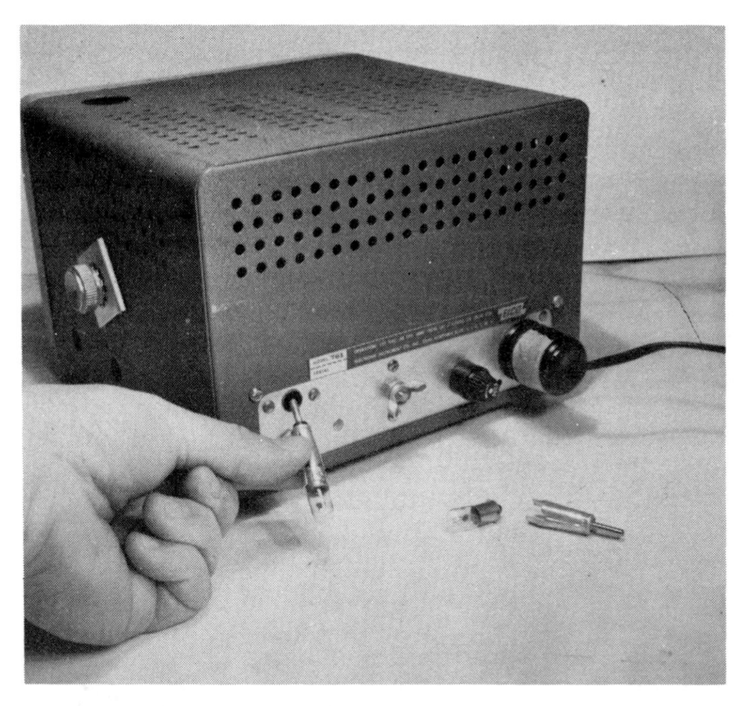

Fig. 9-2. Dummy load made from a No. 47 lamp and antenna plug.

light, and adjustments are made while observing the relative brightness of the pilot lamp. If the transceiver is tuned for maximum output with the dummy load, this does not necessarily mean that once it is removed, and the antenna substituted, maximum power will be absorbed by the antenna. The lamp is a valuable check on transmitter performance, but it is not the proper load which the antenna presents. It is therefore recommended that a final tune-up be accomplished with an swr indicator. With the dummy load connected, voice modulation may be checked —while speaking into the microphone, the operator should see the bulb flicker and increase in brightness with each syllable he utters.

Many transceivers have a small lamp mounted to their front panel marked POWER OUTPUT or some similar designation. This bulb should not be the only indicator used for tune-up purposes, since it often ignites with no antenna connected. If, however, this lamp does not go on when the transmit switch is thrown, it is a good indication of either very poor tune-up or failure of the crystal to oscillate.

Input Power Measurement

The transmitter output in some sets is tuned to the antenna with a multimeter (volt-ohm-milliammeter in Fig. 9-3) and some type of output indicator which reads the

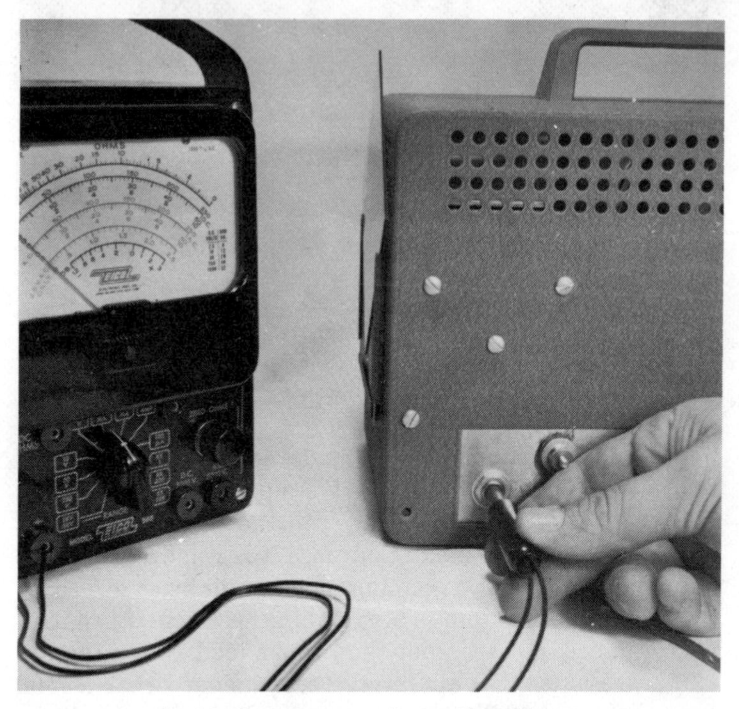

Fig. 9-3. Measuring power input with multimeter.

relative strength of the radio wave emanating from the antenna. The multimeter enables a simple calculation of whether transmitter power *input* is within the 5-watt maximum. Transceivers are generally fitted with a meter jack. First, measure the final rf-amplifier plate voltage at the point indicated in the instruction manual. Next, connect a milliammeter into the circuit and take a current reading. These two values are multiplied, and the result is the power input in watts. While making this calculation, it is important to change milliamperes to amperes; e.g., if the plate voltage is 250 volts dc and the plate current is 18 milliamperes, the input power is 4.5 watts. In working this out, the 18 milliamperes was changed to amperes by moving the decimal three places to the left. Thus:

$$250 \times .018 = 4.5 \text{ watts.}$$

Output Power Measurement

It is valuable to have some form of output indicator for adjusting transmitting output in conjunction with the check just described. You can use a wattmeter or one of the relatively inexpensive devices (Fig. 9-4), described in Chapter 7, permit accurate tuneup. If possible, final

Fig. 9-4. Antenna output meter.

tuning should be done with the transceiver in its mounted position (Fig. 9-5).

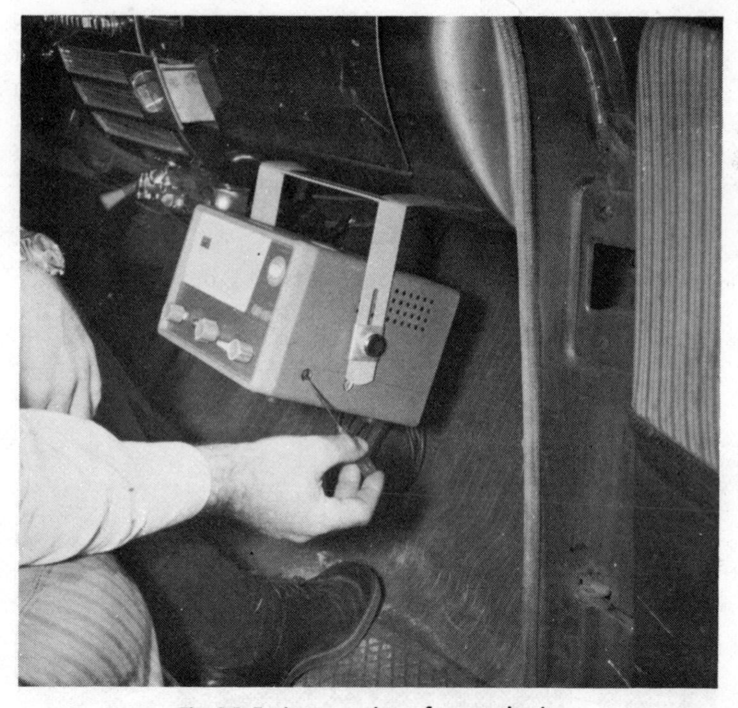

Fig. 9-5. Tuning transmitter of mounted unit.

ANTENNA TUNING

Most antennas for CB are not tunable and are supplied to match the output (52 ohms) of the transceiver. In some cases, antenna elements are equipped with sliding rods to match the 27-MHz band. In these instances, the instruction manual supplied with the antenna has charts with recommended element lengths or a procedure with an swr meter. Beam antennas, often equipped with sliding sections for optimum matching to the transmission line, are factory pretuned. Similarly, the loading coils used with shortened antennas are cut to the necessary number of turns and encased in plastic. If a standard transmission line and antenna are used, there will be no need, in most cases, to adjust the antenna.

To comply with regulations concerning harmonic radiation, many transceivers incorporate a harmonic trap for suppression of the second harmonic on 54 MHz. This is adjusted best with the transceiver close to a tv set which is tuned to Channel 2. The appropriate trimmer is rotated until the least amount of interference appears on the tv screen.

10

How to Operate

Getting the message through on CB with speed and efficiency depends, to a large degree, on operating technique. A single syllable can often replace an entire sentence, and this is certainly an asset during periods of intense interference or poor radio conditions. The jargon of CB is universal—it serves the needs of all who participate on the band. Some aspects of it are official, station identification, for example. Others, like the "10" code, are informal. Each fulfills the requirements of brief, successful communication.

SELECTING A CHANNEL

There are 23 CB channels and you may choose any one for your communications—except 9 and 11. Channel 9 is for emergency and road instructions, as described later in this chapter. Channel 11 is the national calling channel. It enables you to monitor a single frequency, ready to respond to another station calling your identifying numbers. You can then arrange to shift to some open channel and quickly release 11 for calling by others.

Selecting a particular channel is sometimes possible before a transceiver is ordered. Many manufacturers, including kit suppliers, offer the purchaser a choice of crystal frequencies. A common practice is for the user to equip all transceivers which operate in the same network with identical channels. This reduces the amount of receiver tuning and all units simultaneously hear all transmissions.

In multichannel transceivers, where a switch may select one of several frequencies, it may be mandatory to choose crystals which are not spaced more than two or three channels apart. Greater spacing would require retuning of the transmitter each time the crystal switch is thrown.

The choice of a particular channel must remain with the user of the equipment. There is no difference between the transmission characteristics of any of the 23 channels; they all provide the same operating range. If it is possible to monitor the traffic in your area for a day or two, determine which channels are crowded and try to avoid them when choosing your crystal.

There are several examples of CBers using certain channels for special purposes. In many boating areas, for example, Channel 13 is popular among boat owners (Fig. 10-1). Single sideband sets are often found on Channel 16. These choices, however, are unofficial and informal. In 1970, after considerable pressure for one channel for emergency use, the FCC responded by setting aside Channel 9 (27.065 MHz) for the purpose. The change was applauded by volunteers monitoring the band for distress calls. Channel 9 had been the unofficial emergency channel for years but was ruined by interference from other stations. The new regulations were designed to sweep nonessential traffic from the channel.

The primary purpose of Channel 9 is twofold. First, it is reserved for emergency communications involving the immediate safety of life of individuals or the immediate protection of property. The second category includes communications to render assistance to a motorist. To clarify its intent, the FCC gives the following examples of proper use of Channel 9:

Fig. 10-1. Sailboat owner using 5-watt handie-talkie on Channel 13.

"There is a fire in the building on the corner of 6th and Main streets."

"Base to Unit 1, the weather bureau has just issued a thunderstorm warning. Bring the sailboats into port."

"I am out of gas on Interstate 95."

"A tornado sighted six miles north of town."

To illustrate several messages that do not suggest immediate danger to people or property, the FCC gives the following examples which do not qualify for transmission on Channel 9:

"This is Halloween Patrol Unit 3. Everything is quiet here."

"This is observation post 10. No tornado sighted."

"Traffic is moving smoothly on the beltway."

"I am out of gas in my driveway."

When several messages qualify for Channel 9, the FCC suggests priorities. The first, or most urgent, would relate to a dangerous situation in existence, such as a fire or auto accident. Next would be a potentially hazardous situation, like cars stalled at a dangerous location, a lost child, or a boat out of gas. Then there would be messages for helping a disabled motorist and, finally, giving road and street instructions to a motorist. Channel 9 should not be used as a nonemergency calling channel.

In anticipating widespread monitoring of Channel 9, manufacturers designed specialized equipment to make the job easier. The base station in Fig. 10-2, for example,

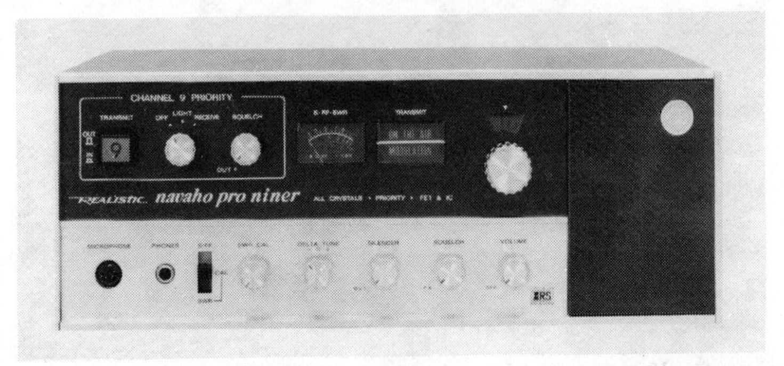

Courtesy Radio Shack

Fig. 10-2. Transceiver base station has controls (upper left) for continuous Channel 9 monitoring.

is a conventional transceiver, except that it automatically signals that a call is arriving on Channel 9, even as the operator listens to a different channel. Another special item (Fig. 10-3) is a receiver-only tuned exclusively to Channel 9. When a call on that channel arrives, the operator is alerted to listen.

PROCEDURE

A prime operating requirement is station identification. This is the transmission of the call letters which appear on the face of the CB radio license. Each of the units covered by a single license is identified by the same call, but

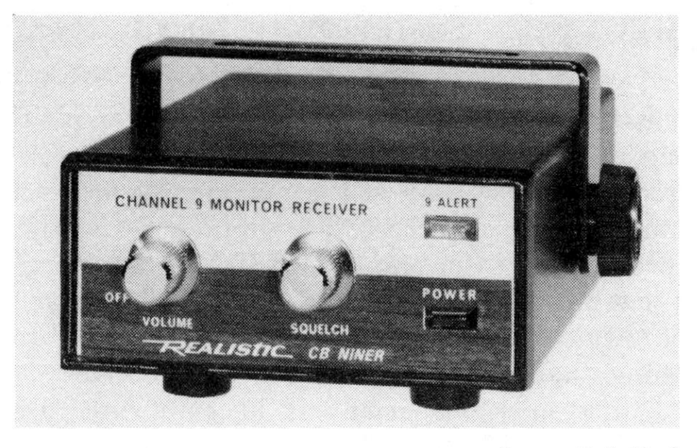

Courtesy Radio Shack

Fig. 10-3. Receiver for monitoring Channel 9 only.

it is possible to distinguish between them by assigning a unit number to each transceiver. The base station might be 00W888, unit 1, with the mobile units identified by 00W888, unit 2; 00W888, unit 3; etc.

The general rule is to give the station identification at the beginning and end of every transmission. During a rapid exchange of words between two units of the same station, this becomes awkward and the rules make allowance for it. In a series of transmissions, the call signs need be transmitted only once every 10 minutes. Assuming that total communication time runs under 10 minutes, call signs need be given only at the open and close of the series. Here is an example of typical operating procedure:

Base Station: "This is 00W888, unit 1, calling unit 2. Over."

Unit 2: "Unit 1, this is 00W888, unit 2. Over."

Base Station: "Did you complete the service call on Maple Street? Over."

Unit 2: "Yes, just a few minor repairs. Shall I return to base? Over."

Base Station: "No, proceed to warehouse for parts pickup. This is 00W888, unit 1, clear with unit 2."

Unit 2: "This is 00W888, unit 2, clear with unit 1."

The use of call signs was limited in this example to the opening and closing transmissions. Call signs were not required within the body of the communication since the transmissions did not exceed 15 minutes.

In exchanging information on CB, it must be remembered that only one person may talk at a time. There is no opportunity to break in and ask the other fellow to repeat something that has been missed until he says "Over." Keeping transmissions brief and uncomplicated assures intelligibility—especially where one station is in motion and may not be receiving each word clearly.

Time Limit

There are additional restrictions on communications. If you are speaking to another station which is not op-

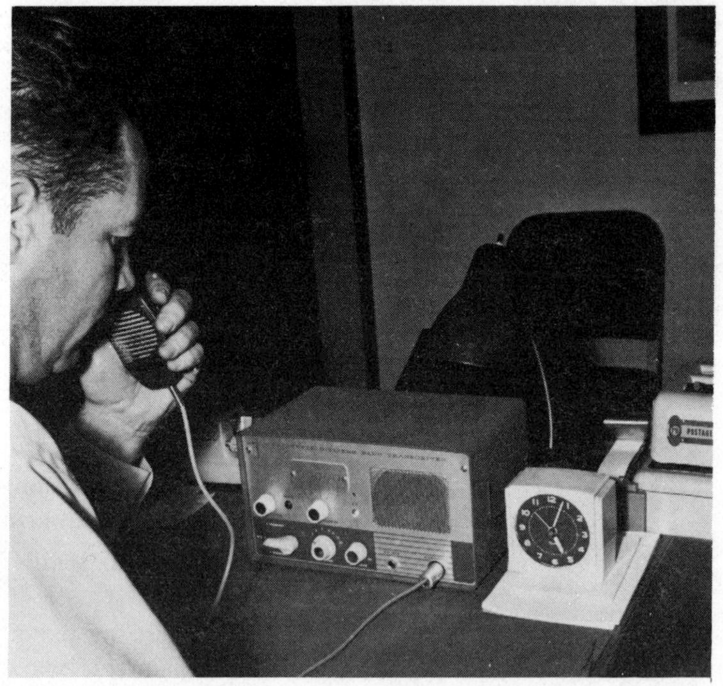

Fig. 10-4. Five-minute rule must be observed during CB operations.

erating under your call sign, this is termed *interstation* communications. By law, you may not speak to such stations for more than five consecutive minutes. When that time elapses, you must observe a one-minute silent period before speaking to that station again. There is, however, no precise time limit if you speak to units of your own system; known as *intrastation* communications, where all stations are under the same license. In this case, you must merely keep length down to the "minimum practicable" time.

"10" Code

Traffic can be speeded up considerably through the use of some form of spoken "abbreviation." Long in use by other radio services, the "10" code serves this need. Numbers take the place of complete ideas, phrases, or sentences and represent word groups which are used repeatedly in the course of daily operation. The consistent use of the number 10 preceding other numbers immediately identifies the use of the code. Operating technique can be sharpened by memorizing and using the excerpts shown in Table 10-1.

Table 10-1. "10" Code

Code Number	Meaning	Code Number	Meaning
10-1	Receiving poorly.	10-18	Engineering test.
10-2	Receiving well.	10-20	What is your location?
10-3	Stop transmitting.	10-21	Call . . . station by phone.
10-4	O.K.	10-23	Stand by.
10-5	Relay message.	10-24	Trouble at station.
10-6	Busy.	10-25	Do you have contact with . . .?
10-7	Out of service—leaving the air.	10-30	Does not conform with rules and regulations.
10-8	In service—subject to call.	10-33	Emergency traffic at this station.
10-9	Repeat, reception bad.	10-36	Correct time.
10-10	Transmission completed, subject to call.	10-65	Clear for message.
10-11	Talking too rapidly.	10-92	Your quality is poor.
10-12	Officials or visitors present.	10-99	Unable to receive your signals.
10-13	Advise weather and road conditions.		

11

Troubleshooting

Citizens Radio equipment, like most other electronic gear, is subject to both sudden and gradual deterioration in performance. In the suggestions outlined in this chapter, it is assumed that a kit-built set is correctly assembled and that there are no wiring errors or poorly soldered joints. In certain instances, a miswired circuit may appear to function well, but may break down after a period of operation—consider any trouble soon after installation with this possibility in mind.

The danger from electrical shock while servicing a transceiver cannot be overemphasized. *Any voltage in excess of 20 volts should be treated as potentially lethal.* This is subject to many conditions such as: the type of circuit, skin resistance of the person, and body resistance to ground. An excellent way of reducing danger is to perform as many steps as possible in the troubleshooting procedure with the power off—this does not mean just with the power switch flipped to the OFF position; if the transceiver operates from house current, there is still one leg of the 117 volts ac connected to a terminal on the switch. *Remove power completely by unplugging the cord from the wall outlet.* Next, hold a screwdriver by its insulated handle and touch the shank to the metal chassis. Then, without removing the shank from the chassis, contact each of the lugs on all large filter capacitors with the screwdriver tip (Fig. 11-1) to drain off any residual charge.

Many tests may be performed with the power off; resistance measurements are the most important. Only if

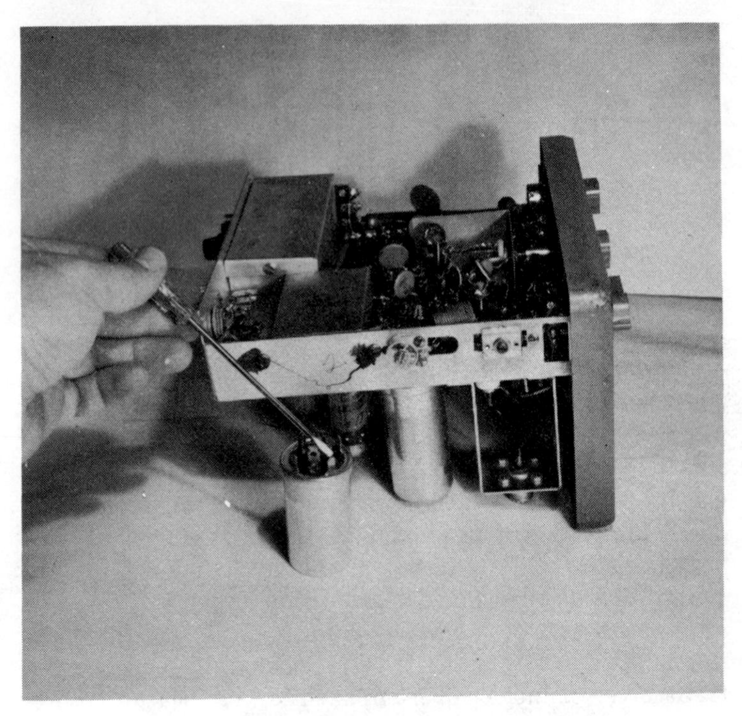

Fig. 11-1. Shorting filter capacitor to prevent shock.

the resistance readings check with the schematic or resistance chart for your set should power be applied for voltage readings. To minimize the effects of accidental contact with voltage, keep one hand away from the chassis while testing. This eliminates at least one path through the body to ground.

TEST EQUIPMENT

For simple troubleshooting a multimeter which can measure voltage, current, and resistance is valuable. Its voltmeter section should have a sensitivity rated at 20,000 ohms per volt. Lower sensitivity will register voltages lower in value than those given in the instruction manual, and, in certain circuits, it can actually short out the voltage you are trying to measure. The vtvm (vacuum-tube voltmeter) affects the circuit least, but unfortunately most have no ammeter provision for measuring current. The ideal is the vom (volt-ohm-milliammeter) which offers the advantage of complete portability; it does not need a power source and operates from a self-contained battery. This is handy for car or marine use where house current is not accessible.

As mentioned earlier, a field-strength meter and wattmeter are desirable. It is a good idea to take output readings while the transceiver is operating properly; then, when trouble is suspected, take another reading and compare the two results. A simple assortment of tools—screwdrivers, nut-drivers (expecially ¼-inch), and small wrenches round out the necessities for basic servicing.

LICENSE REQUIREMENTS FOR SERVICING

Repairs to the frequency-determining section of the transmitter may be made only by, or under the supervision of, persons who hold a First- or Second-Class Radiotelephone license. If the trouble exists in this circuit, the transceiver should be taken to a shop which employs such a licensee. These technicians are not difficult to locate; their license qualifies them to repair all manner of two-way radios—marine, aircraft, police, a-m, fm, tv, or similar transmitting equipment.

LOCALIZING TROUBLE

A frequent cause of faulty equipment can be traced to tubes. Power output falls as tubes lose their emission capability, develop shorted elements, and acquire gas or burned-out filaments. A spare set of tubes should be kept, not only for replacement purposes, but also because they can be used in testing. Substitution is a widely used technique for pinpointing a suspicious tube, and the transceiver itself acts in place of a tube tester.

If tubes are tested on a checker, the mutual conductance type is best. It simulates an operating circuit, subjecting the tube to a more accurate test. On the other hand, the simpler emission tester may not always show negative indications for defective tubes used in high-frequency circuits (as found in a class-D CB radio).

Another troublesome area is the power supply. Its parts often operate at close to maximum ratings, and the heat they develop, especially when ventilation is inadequate, tends to shorten component life below that of other circuits.

An effective method for tracking down trouble is to review the symptoms and isolate the circuit or section in which they appear. This involves some detective work and a block-diagram conception of the transceiver. Once a particular section is singled out as the possible trouble source, the components are checked for their listed value. Narrowing down the circuit in this manner obviates a time-consuming check of every part starting at the antenna jack and ending at the speaker.

Never overlook the possibility of difficulties arising from a mechanical source. It often appears where repeated flexing and manipulation occurs during normal operation, as in microphone cable or power leads. Wires can rupture or short-circuit without any visible indication when hidden by an insulating jacket. Switch contacts lose their springiness and function intermittently. Nuts and bolts loosen after months of operation as vibration takes its toll. Dirt and oxide coatings which form on contacts must be removed (Fig. 11-2) for peak efficiency. Contacts can be cleaned with a strip of paper, a crocus cloth, or with cleaning solution.

Make every effort to diagnose the cause of improper operation while the transceiver is mounted in its normal operating position. Many units have been opened up on the service bench before the defect was found to be produced by something external to the chassis: antenna, power source, broken cable, or unplugged line cord!

If it is decided to remove the unit for closer inspection, the eye and ear can be relied on for initial tests. Careful visual examination reveals components blackened and charred from overload, shorts, and capacitors which have leaked their contents. The ear can perceive the quality and level of the sound issuing from the speaker and can

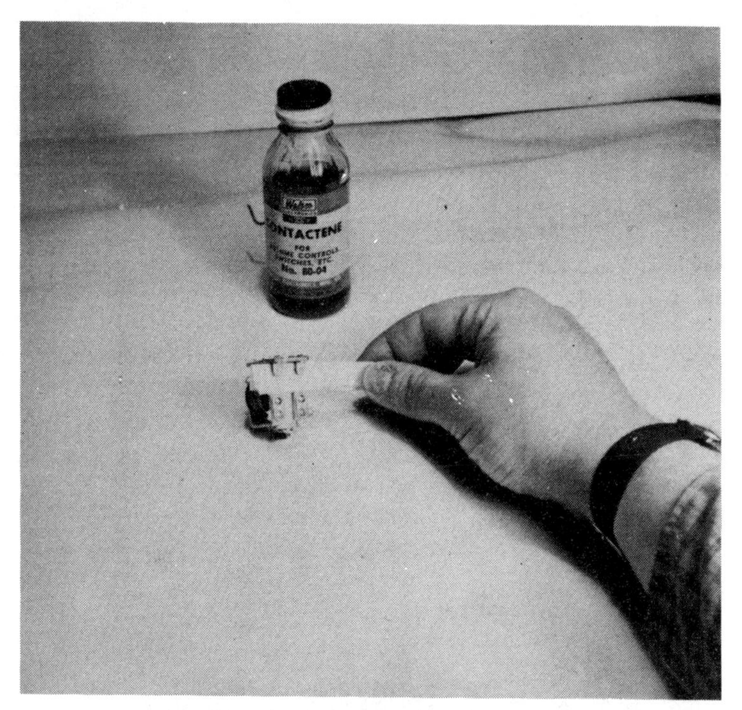

Fig. 11-2. Cleaning send-receive relay contacts.

help locate the arcing of voltage where parts are too close to each other or are partially shorted. Though most trouble is uncovered only by meter indication, these initial tests are worthwhile since they can be done rapidly.

Many cases of unsuccessful servicing can be traced to treating the symptom instead of the cause. A typical example is the replacement of a bad filter resistor in the power supply—only to find that the new one is destroyed within hours or days. Of course, any component may spontaneously fail because it has reached the end of its useful life. But, the question should always be asked, "Is the failure due to some other defective part?" Filter resistors can be overloaded by a short in the filter capacitor which might not display any *visual* sign. Thus, component replacement should always be accompanied by a meter checkout (Fig. 11-3) of the associated circuitry.

SYMPTOM AND CAUSE

The symptoms in this section fall into three major categories: afflictions of the receiver, of the transmitter, and of the two sections simultaneously. The service hints are general enough to apply to most transceiver circuitry, but always refer to your manual for specific values of voltage, resistance, or current. First, difficulties which affect the overall operation of the transceiver.

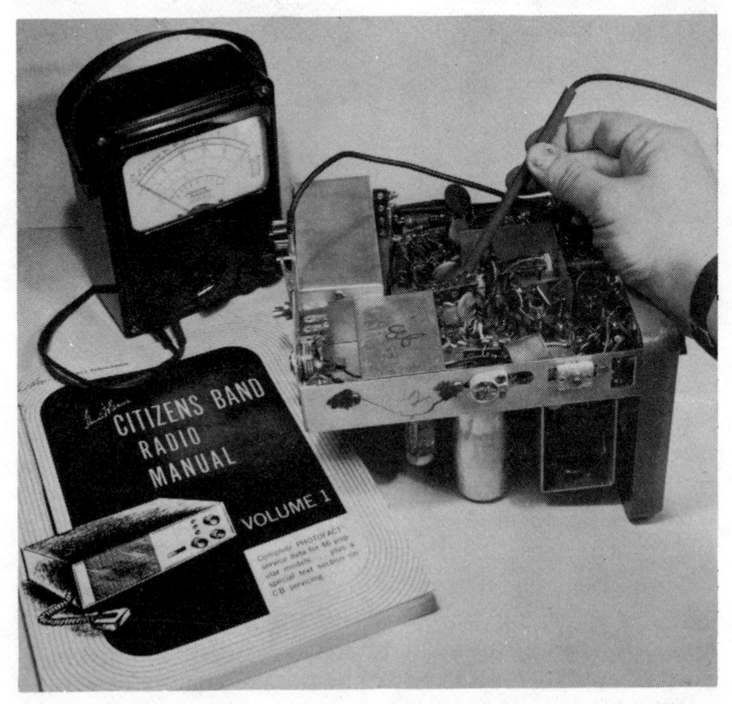

Fig. 11-3. Service literature plus voltage and current measurements will pinpoint most troubles.

Totally Inoperative

This points toward power source or power supply. Check for 117 volts ac if the unit is line operated, or 6 or 12 volts dc if the models use battery power. Is a fuse of correct rating inserted into the fuse receptacle (Fig. 11-4)? An ohms check (power to unit must be discon-

Fig. 11-4. Different fuses are often used for automobile and home operation.

nected) across terminals of the power switch should indicate zero ohms if the switch is on. Determine whether power cables have continuity and all connectors are making firm contact.

Blown Fuses

Short circuits are often in the power supply. Check the wiring for broken insulation. Run an ohms check on the windings of the power transformer. In tube-type ac models the black transformer leads (primary) should read approximately 5 to 30 ohms, red leads (secondary) about 100 ohms, and green filament (secondary) about 1 ohm. A shorted B+ circuit should be checked at its origin—the cathode of the rectifier. The resistance to ground must exceed 15 or 20 thousand ohms. If lower, follow the B+ lead and find the branch which reads the lowest resistance. Disconnect it and check the value of the resistors, capacitors, and other components across the B+ lead,

Fig. 11-5. Replace buffer capacitor when installing new vibrator.

particularly those which go to ground. Shorts in the rectifier tube or filter capacitor are also likely to blow fuses.

In battery-operated power supplies of tube sets, check for a stalled vibrator. If the vibrator is replaced (Fig. 11-5), the buffer capacitor (a unit of 1000 volts or more across the secondary of the power transformer) must be changed at the same time. It can be responsible for vibrator failure.

No Sound In Speaker

Turn volume up to full and place your ear about one inch from the speaker. If a soft hum is heard, this is a good indication that the ac power supply and speaker are functioning and the trouble could be in the audio amplifier. If tubes check good, touch a screwdriver tip to the grid of the tube which feeds the audio output transformer. A definite "click" should be heard in the speaker

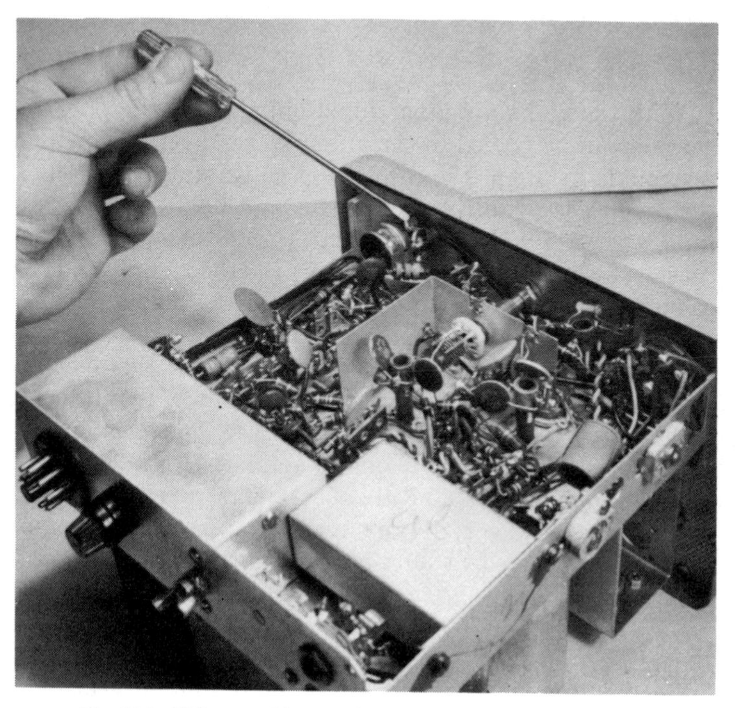

Fig. 11-6. Click test with screwdriver tip applied to control lugs.

if this stage is operating. Repeat this on the audio-driver tube preceding the output tube. The test should not be done to transistor circuits.

The absence of noise as the circuit is disturbed in this manner narrows the range of possible faulty stages. Resistance and voltage checks can also be made to isolate the specific defect.

Try the click test on each of the three lugs which protrude from the volume control (Fig. 11-6). The complete audio amplifier is probably functioning if the disturbance is heard in the speaker when the screwdriver touches the center lug of the control. This suggests that the trouble precedes the amplifier.

It should be remembered that the audio amplifier serves both transmit and receive sections in a CB radio. If your transmitted signals are intelligible, the problem is in the receive section.

Poor Reception

With the audio amplifier functioning properly, poor reception can be traced to the antenna, send-receive switch, or trouble in one of the receiver rf or i-f stages. The click test, which is primitive but useful, can be used to advantage, as in the audio section. Touch each tube grid in succession, working your way toward the front end or antenna-input stage of the receiver. (Note that the detector tube has no grid. Touch its plate.)

If strong atmospheric noise is present in the speaker but no signals are coming through, the local oscillator is to be suspected—defective tube, transistor, resistor, capacitor, coil, or crystal if one is employed.

The transformer and coil adjustments rarely go out of alignment. However, if any of their associated components are replaced, a slight realignment might be necessary.

Low Transmitter Output

Correct tuning of the transmitter output is a critical point and should be checked. A fraction of a turn on any of its adjustments can cut power to unusable levels. Any time the antenna is changed, a new tune-up is in order. Check the antenna and its cables. These factors will affect the transmitter more adversely than they will affect the receiver.

Other causes of poor transmitter performance are commonly traceable to the frequency-determining section, namely the oscillator stage. Complaints of frequency drift or sluggish response when the SEND-RECEIVE switch is thrown to SEND can be produced by an aging crystal.

The flicker of the output indicator light, on the rf meter, on the front panel of many transceivers gives evidence of an oscillating crystal and an approximate indication of modulation.

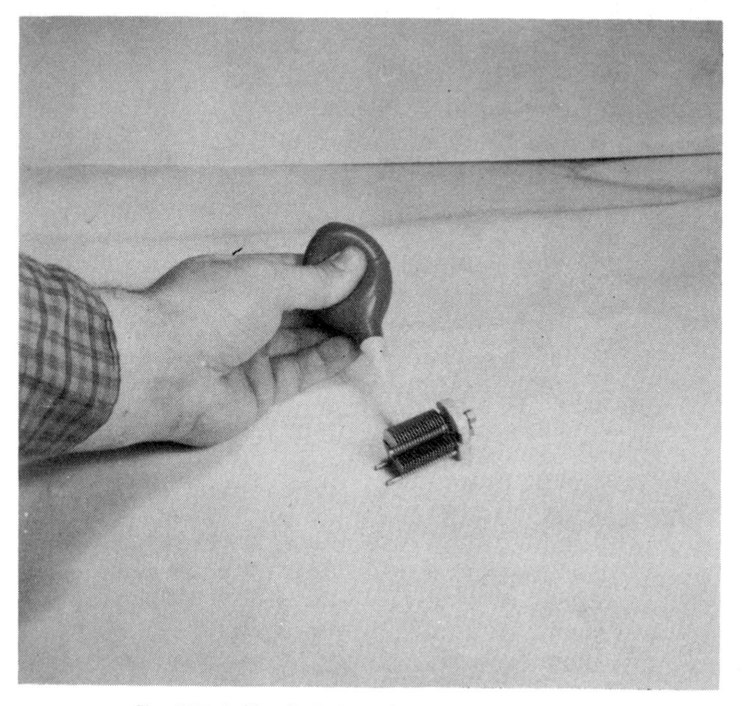

Fig. 11-7. Rubber bulb is useful for blowing out dirt.

PREVENTING TROUBLE

A program of preventive maintenance will do much to forestall trouble. This is important if the CB serves in business applications where equipment downtime can be expensive. There can be more serious implications, too; e.g., a small boat in rough waters with a dead transceiver. Periodic maintenance should include the following: a tube check, close visual inspection, and performance tests under identical conditions each time. Also clean and lubricate all switches and sliding contacts with chemicals manufactured for the purpose. Blow out dust which collects inside the cabinet—especially between the moving plates of the tuning capacitor (Fig. 11-7). Dissolve any grease or sludge which deposits on or near components. Use crocus cloth on plugs and relay contacts to restore their original shiny surfaces. Tighten nuts and bolts, tubes in their sockets, and panel knobs. Replace burned-out pilot lamps. Wipe away soot or other formations on antenna terminals and elements.

12

The FCC and You

To acquire a CB license, you must fulfill certain requirements: citizenship, age, and an acceptable use of the band. And, as you operate, the responsibilities continue. The benefits which accrue from operating a two-way radio exist only as long as the rules are observed. Many of these rules appear in detail in Part 95 of the regulations—others are recommended to forestall further FCC restrictions.

PERMISSIBLE COMMUNICATIONS

The messages sent out over CB are primarily for communication between units of the same station. Other stations may be contacted only for substantive messages relating to the personal or business activities of the individuals concerned. Within narrow limits, there are certain interpretations of these rulings. One example would be when you are driving in an unfamiliar part of the country looking for a motel. In this instance you may transmit an inquiry on Channel 9. This is in sharp contrast to sending a *CQ,* or general call, to raise an unknown station for the sake of conversation or amusement.

*DX'*ing (to see how far your signals reach) is also forbidden. To remain within the category of a short-range operator, transmissions must be directed toward specific stations not more than 150 miles distant.

Transmitting music or charging a fee for your communication services is taboo. One-way transmission is allowed if the messages are addressed to specific units. Hooking the output of the receiver into a public address system for broadcasting to the public is also forbidden.

STATION LICENSE REQUIREMENTS

Licenses are issued for a term of five years. Near the end of this period, an FCC Form No. 505 (identical to the initial application) is filed for renewal. This extends the license for another five years. Unlike other radio call assignments, your CB call letters will be different each five-year period. This system was adopted to simplify FCC bookkeeping. Take this into consideration if you plan to have your call printed on stationery or other literature.

If modifications to the license are made, not only does the form No. 505 have to be submitted again, but in all probability the call letters also will change when the amended license returns to you. Anytime you vary the number of units in your network, change address, or area of communication, a new form is to be submitted for approval.

There is another FCC form, rarely needed by class-D licensees, dealing with antenna structures. This is form No. 401-A which might be required for a proposed antenna which exceeds an overall height of 170 feet above ground level or is located close to an airport. But the basic rule that class-D antennas should not exceed 20 feet in height over an existing structure will probably cover most cases.

Licenses must be posted or readily visible for inspection. Many transceivers provide a transparent pocket on the side or bottom of the cabinet for this purpose (Fig. 12-1). Since a single license is issued for several units, transmitter identification cards (FCC Form No. 452-C) can be used for the additional sets. Some manufacturers provide these cards with their equipment. Each card should bear the call sign and licensee's name and address.

TYPICAL VIOLATIONS

While CB is intended for private communications, it is by no means a private system; anyone may listen, if he is

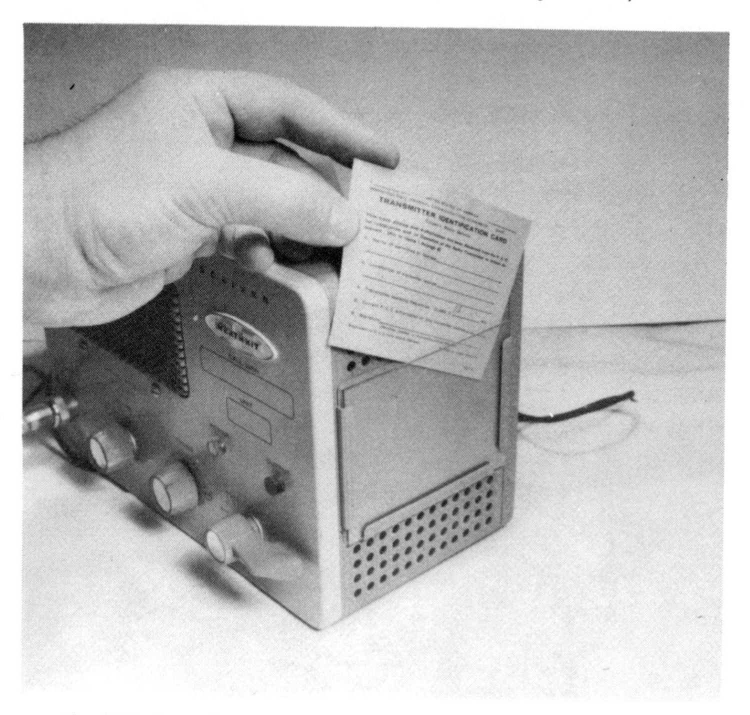

Fig. 12-1. Transmitter identification card must be attached to all units.

within range of your signals. With proper receiving equipment and a high-gain antenna, your signals can often be picked up many miles beyond your normal service area. And this is done by FCC monitoring points (Fig. 12-2). The result is an issuance of citations to violators of the rules and regulations. These notices must be answered within 10 days of receipt with a written statement explaining the matter. It also must state the steps taken to prevent future violations of the type cited. Depending on the nature of the violation, three courses of action may follow—the notice goes into FCC files for future reference, suspension of the license, or revocation of the license.

Two major categories appear to account for most citations: off-frequency operation and improper use of the band. The frequency of the transmitter should be periodi-

Fig. 12-2. FCC monitoring stations are aided by mobile units.

cally checked to discover circuit defects which might cause excessive deviation from the nominal frequency (Fig. 12-3) of the crystal. Faulty tune-up of the crystal oscillator can produce similar difficulty.

Improper Use

The large number of citations for improper use of CB, according to one FCC bulletin, is partly due to misunderstanding regarding the band. Since the frequencies were formerly associated with amateur radio, some operators continue to use the band in a similar manner. The bulletin also states that many who could not or would not obtain ham licenses seized upon the class-D service to carry out activities common to amateur radio. The *CQ,* "rag chew,"

Fig. 12-3. Exchanging crystals between units of different manufacturers might cause off-frequency operation.

and the fierce competition for *DX* which make ham radio a dynamic medium for the hobbyist or experimenter reduces the utility of the Citizens band. It is surprising that persons who wish to use two-way radio as a source of diversion and electronic interest do not secure the Novice Class amateur license. The examination is modest in its requirements and the frequency allocations are far more liberal than the 23 fixed channels of the class-D band.

Appendix

FCC FIELD OFFICES

Mailing addresses for Commission Field Offices are listed below. Street addresses can be found in local directories under the United States Government listings. Address all communications to Engineer in Charge.

FIELD ENGINEERING OFFICES

Alabama, Mobile 36602
Alaska, Anchorage 99510 (P.O. Box 644)
California, Los Angeles 90012
California, San Diego 92101
California, San Francisco 94111
Colorado, Denver 80202
District of Columbia, Washington 20554
Florida, Miami 33130
Florida, Tampa 33602
Georgia, Atlanta 30303
Georgia, Savannah 31402 (P.O. Box 8004)
Hawaii, Honolulu 96808 (P.O. Box 1021)
Illinois, Chicago 60604

Louisiana, New Orleans 70130
Maryland, Baltimore 21201
Massachusetts, Boston 02109
Michigan, Detroit 48226
Minnesota, St. Paul 55101
Missouri, Kansas City 64106
New York, Buffalo 14203
New York, New York 10014
Oregon, Portland 97204
Pennsylvania, Philadelphia 19106
Puerto Rico, San Juan 00903 (P.O. Box 2987)
Texas, Beaumont 77701
Texas, Dallas 75202
Texas, Houston 77002
Virginia, Norfolk 23510
Washington, Seattle 98104

Index